# ZooKeeper

*Flavio Junqueira and Benjamin Reed*

Beijing • Cambridge • Farnham • Köln • Sebastopol • Tokyo

**SHROFF PUBLISHERS & DISTRIBUTORS PVT. LTD.**

*Mumbai* *Bangalore* *Kolkata* *New Delhi*

## ZooKeeper

by Flavio Junqueira and Benjamin Reed

 ISBN: 978-1-449-36130-3
Originally printed in the United States of America.

Published by O'Reilly Media, Inc., 1005 Gravenstein Highway North, Sebastopol, CA 95472.

O'Reilly books may be purchased for educational, business, or sales promotional use. Online editions are also available for most titles (*http://safari.oreilly.com*). For more information, contact our corporate/institutional sales department: (800) 998-9938 or *corporate@oreilly.com*.

**Editors:** Mike Loukides and Andy Oram
**Production Editor:** Kara Ebrahim
**Copyeditor:** Kim Cofer
**Proofreader:** Rachel Head
**Indexer:** Judy McConville
**Cover Designer:** Randy Comer
**Interior Designer:** David Futato
**Illustrator:** Rebecca Demarest

**Printing History:**

November 2013: First Edition

**Revision History for the First Edition:** 2013-11-15: First release

*See http://oreilly.com/catalog/errata.csp?isbn=9781449361303 for release details.*

**Third Indian Reprint:** June 2017

ISBN : 978-93-5110-834-4



Published by **Shroff Publishers and Distributors Pvt. Ltd.** C-103, TTC Industrial Area, MIDC, Pawane, Navi Mumbai - 400 703 • TEL: (91 22) 4158 4158 • FAX: (91 22) 4158 4141• E-mail : spdorders@shroffpublishers.com Web : www.shroffpublishers.com Printed at Jasmine Art Printers Pvt. Ltd., Navi Mumbai.

# Table of Contents

## Part II. Programming with ZooKeeper

# Part III. Administering ZooKeeper

# Preface

Building distributed systems is hard. A lot of the applications people use daily, however, depend on such systems, and it doesn't look like we will stop relying on distributed computer systems any time soon. Apache ZooKeeper has been designed to mitigate the task of building robust distributed systems. It has been built around core distributed computing concepts, with its main goal to present the developer with an interface that is simple to understand and program against, thus simplifying the task of building such systems.

Even with ZooKeeper, the task is not trivial—which leads us to this book. This book will get you up to speed on building distributed systems with Apache ZooKeeper. We start with basic concepts that will quickly make you feel like you're a distributed systems expert. Perhaps it will be a bit disappointing to see that it is not that simple when we discuss a bunch of caveats that you need to be aware of. But don't worry; if you understand well the key issues we expose, you'll be on the right track to building great distributed applications.

## Audience

This book is aimed at developers of distributed systems and administrators of applications using ZooKeeper in production. We assume knowledge of Java, and try to give you enough background in the principles of distributed systems to use ZooKeeper robustly.

## Contents of This Book

Part I covers some motivations for a system like Apache ZooKeeper, and some of the necessary background in distributed systems that you need to use it.

- Chapter 1, *Introduction*, explains what ZooKeeper can accomplish and how its design supports its mission.

- Chapter 2, *Getting to Grips with ZooKeeper*, goes over the basic concepts and building blocks. It explains how to get a more concrete idea of what ZooKeeper can do by using the command line.

Part II covers the library calls and programming techniques that programmers need to know. It is useful but not required reading for system administrators. This part focuses on the Java API because it is the most popular. If you are using a different language, you can read this part to learn the basic techniques and functions, then implement them in a different language. We have an additional chapter covering the C binding for the developers of applications in this language.

- Chapter 3, *Getting Started with the ZooKeeper API*, introduces the Java API.
- Chapter 4, *Dealing with State Change*, explains how to track and react to changes to the state of ZooKeeper.
- Chapter 5, *Dealing with Failure*, shows how to recover from system or network problems.
- Chapter 6, *ZooKeeper Caveat Emptor*, describes some miscellaneous but important considerations you should look for to avoid problems.
- Chapter 7, *The C Client*, introduces the C API, which is the basis for all the non-Java implementations of the ZooKeeper API. Therefore, it's valuable for programmers using any language besides Java.
- Chapter 8, *Curator: A High-Level API for ZooKeeper*, describes a popular high-level interfaces to ZooKeeper.

Part III covers ZooKeeper for system administrators. Programmers might also find it useful, in particular the chapter about internals.

- Chapter 9, *ZooKeeper Internals*, describes some of the choices made by ZooKeeper developers that have an impact on administration tasks.
- Chapter 10, *Running ZooKeeper*, shows how to configure ZooKeeper.

## Conventions Used in this Book

The following typographical conventions are used in this book:

*Italic*
: Used for emphasis, new terms, URLs, commands and utilities, and file and directory names.

`Constant width`
: Indicates variables, functions, types, parameters, objects, and other programming constructs.

**`Constant width bold`**
: Shows commands or other text that should be typed literally by the user. Also used for emphasis in command output.

*`Constant width italic`*
: Indicates placeholders in code or commands that should be replaced by appropriate values.

This icon signifies a tip, suggestion, or a general note.

## Using Code Examples

Supplemental material (code examples, exercises, etc.) is available for download at *http://bit.ly/zookeeper-code*.

This book is here to help you get your job done. In general, if example code is offered with this book, you may use it in your programs and documentation. You do not need to contact us for permission unless you're reproducing a significant portion of the code. For example, writing a program that uses several chunks of code from this book does not require permission. Selling or distributing a CD-ROM of examples from O'Reilly books does require permission. Answering a question by citing this book and quoting example code does not require permission. Incorporating a significant amount of example code from this book into your product's documentation does require permission.

We appreciate, but do not require, attribution. An attribution usually includes the title, author, publisher, and ISBN. For example: "*ZooKeeper* by Flavio Junqueira and Benjamin Reed (O'Reilly). Copyright 2014 Flavio Junqueira and Benjamin Reed, 978-1-449-36130-3."

If you feel your use of code examples falls outside fair use or the permission given above, feel free to contact us at *permissions@oreilly.com*.

## Safari® Books Online

*Safari Books Online* is an on-demand digital library that delivers expert content in both book and video form from the world's leading authors in technology and business.

Technology professionals, software developers, web designers, and business and creative professionals use Safari Books Online as their primary resource for research, problem solving, learning, and certification training.

Safari Books Online offers a range of product mixes and pricing programs for organizations, government agencies, and individuals. Subscribers have access to thousands of books, training videos, and prepublication manuscripts in one fully searchable database from publishers like O'Reilly Media, Prentice Hall Professional, Addison-Wesley Professional, Microsoft Press, Sams, Que, Peachpit Press, Focal Press, Cisco Press, John Wiley & Sons, Syngress, Morgan Kaufmann, IBM Redbooks, Packt, Adobe Press, FT Press, Apress, Manning, New Riders, McGraw-Hill, Jones & Bartlett, Course Technology, and dozens more. For more information about Safari Books Online, please visit us online.

## How to Contact Us

Please address comments and questions concerning this book to the publisher:


O'Reilly Media, Inc.<br>
1005 Gravenstein Highway North<br>
Sebastopol, CA 95472<br>
800-998-9938 (in the United States or Canada)<br>
707-829-0515 (international or local)<br>
707-829-0104 (fax)


We have a web page for this book, where we list errata, examples, and any additional information. You can access this page at *http://oreil.ly/zookeeper-orm*.

To comment or ask technical questions about this book, send email to *bookquestions@oreilly.com*.

For more information about our books, courses, conferences, and news, see our website at *http://www.oreilly.com*.

Find us on Facebook: *http://facebook.com/oreilly*

Follow us on Twitter: *http://twitter.com/oreillymedia*

Watch us on YouTube: *http://www.youtube.com/oreillymedia*

## Acknowledgments


We would like to thank our editors, initially Nathan Jepson and later Andy Oram, for the fantastic job they did of getting us to produce this book.

We would like to thank our families and employers for understanding the importance of spending so many hours with this book. We hope you appreciate the outcome.

We would like to thank our reviewers for spending time to give us great comments that helped us to improve the material in this book. They are: Patrick Hunt, Jordan Zimmerman, Donald Miner, Henry Robinson, Isabel Drost-Fromm, and Thawan Kooburat.

ZooKeeper is the collective work of the Apache ZooKeeper community. We work with some really excellent committers and other contributors; it's a privilege to work with you all. We also want to give a big thanks to all of the ZooKeeper users who have reported bugs and given us so much feedback and encouragement over the years.

# PART I
# ZooKeeper Concepts and Basics

This part of the book should be read by anyone interested in ZooKeeper. It explains the problems that ZooKeeper solves and the trade-offs made during its design.

# CHAPTER 1
# Introduction

In the past, each application was a single program running on a single computer with a single CPU. Today, things have changed. In the Big Data and Cloud Computing world, applications are made up of many independent programs running on an ever-changing set of computers.

Coordinating the actions of these independent programs is far more difficult than writing a single program to run on a single computer. It is easy for developers to get mired in coordination logic and lack the time to write their application logic properly—or perhaps the converse, to spend little time with the coordination logic and simply to write a quick-and-dirty master coordinator that is fragile and becomes an unreliable single point of failure.

ZooKeeper was designed to be a robust service that enables application developers to focus mainly on their application logic rather than coordination. It exposes a simple API, inspired by the filesystem API, that allows developers to implement common coordination tasks, such as electing a master server, managing group membership, and managing metadata. ZooKeeper is an application library with two principal implementations of the APIs—Java and C—and a service component implemented in Java that runs on an ensemble of dedicated servers. Having an ensemble of servers enables ZooKeeper to tolerate faults and scale throughput.

When designing an application with ZooKeeper, one ideally separates application data from control or coordination data. For example, the users of a web-mail service are interested in their mailbox content, but not on which server is handling the requests of a particular mailbox. The mailbox content is application data, whereas the mapping of the mailbox to a specific mail server is part of the coordination data (or metadata). A ZooKeeper ensemble manages the latter.

## The ZooKeeper Mission

Trying to explain what ZooKeeper does for us is like trying to explain what a screwdriver can do for us. In very basic terms, a screwdriver allows us to turn or drive screws, but putting it this way does not really express the power of the tool. It enables us to assemble pieces of furniture and electronic devices, and in some cases hang pictures on the wall. By giving some examples like this, we can give a sense of what can be done, but it is certainly not exhaustive.

The argument for what a system like ZooKeeper can do for us is along the same lines: it enables coordination tasks for distributed systems. A coordination task is a task involving multiple processes. Such a task can be for the purposes of cooperation or to regulate contention. Cooperation means that processes need to do something together, and processes take action to enable other processes to make progress. For example, in typical master-worker architectures, the worker informs the master that it is available to do work. The master consequently assigns tasks to the worker. Contention is different: it refers to situations in which two processes cannot make progress concurrently, so one must wait for the other. Using the same master-worker example, we really want to have a single master, but multiple processes may try to become the master. The multiple processes consequently need to implement *mutual exclusion*. We can actually think of the task of acquiring mastership as the one of acquiring a lock: the process that acquires the mastership lock exercises the role of master.

If you have any experience with multithreaded programs, you will recognize that there are a lot of similar problems. In fact, having a number of processes running in the same computer or across computers is conceptually not different at all. Synchronization primitives that are useful in the context of multiple threads are also useful in the context of distributed systems. One important difference, however, stems from the fact that different computers do not share anything other than the network in a typical shared-nothing architecture. While there are a number of message-passing algorithms to implement synchronization primitives, it is typically much easier to rely upon a component that provides a shared store with some special ordering properties, like ZooKeeper does.

Coordination does not always take the form of synchronization primitives like leader election or locks. Configuration metadata is often used as a way for a process to convey what others should be doing. For example, in a master-worker system, workers need to know the tasks that have been assigned to them, and this information must be available even if the master crashes.

Let's look at some examples where ZooKeeper has been useful to get a better sense of where it is applicable:

*Apache HBase*
: HBase is a data store typically used alongside Hadoop. In HBase, ZooKeeper is used to elect a cluster master, to keep track of available servers, and to keep cluster metadata.

*Apache Kafka*
: Kafka is a pub-sub messaging system. It uses ZooKeeper to detect crashes, to implement topic discovery, and to maintain production and consumption state for topics.

*Apache Solr*
: Solr is an enterprise search platform. In its distributed form, called SolrCloud, it uses ZooKeeper to store metadata about the cluster and coordinate the updates to this metadata.

*Yahoo! Fetching Service*
: Part of a crawler implementation, the Fetching Service fetches web pages efficiently by caching content while making sure that web server policies, such as those in *robots.txt* files, are preserved. This service uses ZooKeeper for tasks such as master election, crash detection, and metadata storage.

*Facebook Messages*
: This is a Facebook application that integrates communication channels: email, SMS, Facebook Chat, and the existing Facebook Inbox. It uses ZooKeeper as a controller for implementing sharding and failover, and also for service discovery.

There are a lot more examples out there; this is a just a sample. Given this sample, let's now bring the discussion to a more abstract level. When programming with ZooKeeper, developers design their applications as a set of clients that connect to ZooKeeper servers and invoke operations on them through the ZooKeeper client API. Among the strengths of the ZooKeeper API, it provides:

- Strong consistency, ordering, and durability guarantees
- The ability to implement typical synchronization primitives
- A simpler way to deal with many aspects of concurrency that often lead to incorrect behavior in real distributed systems

ZooKeeper, however, is not magic; it will not solve all problems out of the box. It is important to understand what ZooKeeper provides and to be aware of its tricky aspects. One of the goals of this book is to discuss ways to deal with these issues. We cover the basic material needed to get the reader to understand what ZooKeeper actually does for developers. We additionally discuss several issues we have come across while implementing applications with ZooKeeper and helping developers new to ZooKeeper.

### The Origin of the Name "ZooKeeper"

ZooKeeper was developed at Yahoo! Research. We had been working on ZooKeeper for a while and pitching it to other groups, so we needed a name. At the time the group had been working with the Hadoop team and had started a variety of projects with the names of animals, Apache Pig being the most well known. As we were talking about different possible names, one of the group members mentioned that we should avoid another animal name because our manager thought it was starting to sound like we lived in a zoo. That is when it clicked: distributed systems *are* a zoo. They are chaotic and hard to manage, and ZooKeeper is meant to keep them under control.

The cat on the book cover is also appropriate, because an early article from Yahoo! Research about ZooKeeper described distributed process management as similar to herding cats. ZooKeeper sounds much better than CatHerder, though.

## How the World Survived without ZooKeeper

Has ZooKeeper enabled a whole new class of applications to be developed? That doesn't seem to be the case. ZooKeeper instead simplifies the development process, making it more agile and enabling more robust implementations.

Previous systems have implemented components like distributed lock managers or have used distributed databases for coordination. ZooKeeper, in fact, borrows a number of concepts from these prior systems. It does not expose a lock interface or a general-purpose interface for storing data, however. The design of ZooKeeper is specialized and very focused on coordination tasks. At the same time, it does not try to impose a particular set of synchronization primitives upon the developer, being very flexible with respect to what can be implemented.

It is certainly possible to build distributed systems without using ZooKeeper. ZooKeeper, however, offers developers the possibility of focusing more on application logic rather than on arcane distributed systems concepts. Programming distributed systems without ZooKeeper is possible, but more difficult.

## What ZooKeeper Doesn't Do

The ensemble of ZooKeeper servers manages critical application data related to coordination. ZooKeeper is not for bulk storage. For bulk storage of application data, there are a number of options available, such as databases and distributed file systems. When designing an application with ZooKeeper, one ideally separates application data from control or coordination data. They often have different requirements; for example, with respect to consistency and durability.

ZooKeeper implements a core set of operations that enable the implementation of tasks that are common to many distributed applications. How many applications do you know that have a master or need to track which processes are responsive? ZooKeeper, however, does not implement the tasks for you. It does not elect a master or track live processes for the application out of the box. Instead, it provides the tools for implementing such tasks. The developer decides what coordination tasks to implement.

## The Apache Project

ZooKeeper is an open source project hosted by the Apache Software Foundation. It has a Project Management Committee (PMC) that is responsible for management and oversight of the project. Only *committers* can check in patches, but any developer can contribute a patch. Developers can become committers after contributing to the project. Contributions to the project are not limited to patches—they can come in other forms and interactions with other members of the community. We have lots of discussions on the mailing lists about new features, questions from new users, etc. We highly encourage developers interested in participating in the community to subscribe to the mailing lists and participate in the discussions. You may well find it also worth becoming a committer if you want to have a long-term relationship with ZooKeeper through some project.

## Building Distributed Systems with ZooKeeper

There are multiple definitions of a *distributed system*, but for the purposes of this book, we define it as a system comprised of multiple software components running independently and concurrently across multiple physical machines. There are a number of reasons to design a system in a distributed manner. A distributed system is capable of exploiting the capacity of multiple processors by running components, perhaps replicated, in parallel. A system might be distributed geographically for strategic reasons, such as the presence of servers in multiple locations participating in a single application.

Having a separate coordination component has a couple of important advantages. First, it allows the component to be designed and implemented independently. Such an independent component can be shared across many applications. Second, it enables a system architect to reason more easily about the coordination aspect, which is not trivial (as this book tries to expose). Finally, it enables a system to run and manage the coordination component separately. Running such a component separately simplifies the task of solving issues in production.

Software components run in operating system processes, in many cases executing multiple threads. Thus, ZooKeeper servers and clients are processes. Often, a single physical server (whether a standalone machine or an operating system in a virtual environment) runs a single application process, although the process might execute multiple threads to exploit the multicore capacity of modern processors.

Processes in a distributed system have two broad options for communication: they can exchange messages directly through a network, or read and write to some shared storage. ZooKeeper uses the shared storage model to let applications implement coordination and synchronization primitives. But shared storage itself requires network communication between the processes and the storage. It is important to stress the role of network communication because it is an important source of complications in the design of a distributed system.

In real systems, it is important to watch out for the following issues:

*Message delays*
: Messages can get arbitrarily delayed; for instance, due to network congestion. Such arbitrary delays may introduce undesirable situations. For example, process *P* may send a message before another process *Q* sends its message, according to a reference clock, but *Q*'s message might be delivered first.

*Processor speed*
: Operating system scheduling and overload might induce arbitrary delays in message processing. When one process sends a message to another, the overall latency of this message is roughly the sum of the processing time on the sender, the transmission time, and the processing time on the receiver. If the sending or receiving process requires time to be scheduled for processing, then the message latency is higher.

*Clock drift*
: It is not uncommon to find systems that use some notion of time, such as when determining the time at which events occur in the system. Processor clocks are not reliable and can arbitrarily drift away from each other. Consequently, relying upon processor clocks might lead to incorrect decisions.

One important consequence of these issues is that it is very hard in practice to tell if a process has crashed or if any of these factors is introducing some arbitrary delay. Not receiving a message from a process could mean that it has crashed, that the network is delaying its latest message arbitrarily, that there is something delaying the process, or that the process clock is drifting away. A system in which such a distinction can't be made is said to be *asynchronous*.

Data centers are generally built using large batches of mostly uniform hardware. But even in data centers, we have observed the impact of all these issues on applications due to the use of multiple generations of hardware in a single application, and subtle but significant performance differences even within the same batch of hardware. All these things complicate the life of a distributed systems designer.

ZooKeeper has been designed precisely to make it simpler to deal with these issues. ZooKeeper does not make the problems disappear or render them completely transparent to applications, but it does make the problems more tractable. ZooKeeper implements solutions to important distributed computing problems and packages up these implementations in a way that is intuitive to developers… at least, this has been our hope all along.

## Example: Master-Worker Application

We have talked about distributed systems in the abstract, but it is now time to make it a bit more concrete. Let's consider a common architecture that has been used extensively in the design of distributed systems: a master-worker architecture (Figure 1-1). One important example of a system following this architecture is HBase, a clone of Google's Bigtable. At a very high level, the master server (HMaster) is responsible for keeping track of the region servers (HRegionServer) available and assigning regions to servers. Because we don't cover it here, we encourage you to check the HBase documentation for further details on how it uses ZooKeeper. Our discussion instead focuses on a generic master-worker architecture.

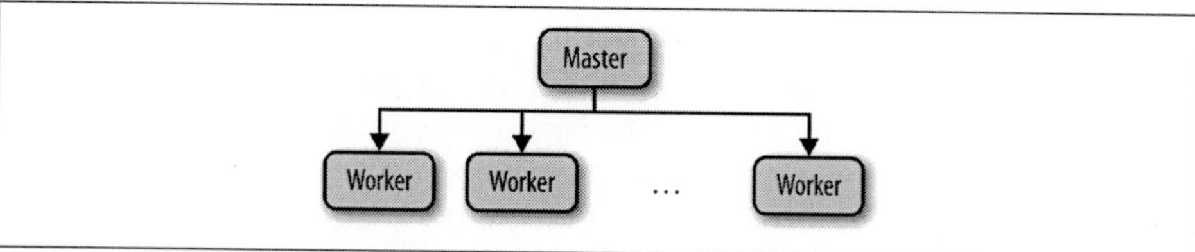


*Figure 1-1. Master-worker example*

In general, in such an architecture a master process is responsible for keeping track of the workers and tasks available, and for assigning tasks to workers. For ZooKeeper, this architecture style is representative because it illustrates a number of popular tasks, like electing a master, keeping track of available workers, and maintaining application metadata.

To implement a master-worker system, we must solve three key problems:

*Master crashes*
: If the master is faulty and becomes unavailable, the system cannot allocate new tasks or reallocate tasks from workers that have also failed.

*Worker crashes*
: If a worker crashes, the tasks assigned to it will not be completed.

*Communication failures*
: If the master and a worker cannot exchange messages, the worker might not learn of new tasks assigned to it.

To deal with these problems, the system must be able to reliably elect a new master if the previous one is faulty, determine which workers are available, and decide when the state of a worker is stale with respect to the rest of the system. We'll look at each task briefly in the following sections.

## Master Failures

To mask master crashes, we need to have a backup master. When the primary master crashes, the backup master takes over the role of primary master. Failing over, however, is not as simple as starting to process requests that come in to the master. The new primary master must be able to recover the state of the system at the time the old primary master crashed. For recoverability of the master state, we can't rely on pulling it from the faulty master because it has crashed; we need to have it somewhere else. This somewhere else is ZooKeeper.

Recovering the state is not the only important issue. Suppose that the primary master is up, but the backup master suspects that the primary master has crashed. This false suspicion could happen because, for example, the primary master is heavily loaded and its messages are being delayed arbitrarily (see the discussion in "Building Distributed Systems with ZooKeeper" on page 7). The backup master will execute all necessary procedures to take over the role of primary master and may eventually start executing the role of primary master, becoming a second primary master. Even worse, if some workers can't communicate with the primary master, say because of a network partition, they may end up following the second primary master. This scenario leads to a problem commonly called *split-brain*: two or more parts of the system make progress independently, leading to inconsistent behavior. As part of coming up with a way to cope with master failures, it is critical that we avoid split-brain scenarios.

## Worker Failures

Clients submit tasks to the master, which assigns the tasks to available workers. The workers receive assigned tasks and report the status of the execution once these tasks have been executed. The master next informs the clients of the results of the execution.

If a worker crashes, all tasks that were assigned to it and not completed must be reassigned. The first requirement here is to give the master the ability to detect worker crashes. The master must be able to detect when a worker crashes and must be able to determine what other workers are available to execute its tasks. In the case a worker crashes, it may end up partially executing tasks or even fully executing tasks but not reporting the results. If the computation has side effects, some recovery procedure might be necessary to clean up the state.

## Communication Failures

If a worker becomes disconnected from the master, say due to a network partition, reassigning a task could lead to two workers executing the same task. If executing a task more than once is acceptable, we can reassign without verifying whether the first worker has executed the task. If it is not acceptable, then the application must be able to accommodate the possibility that multiple workers may end up trying to execute the task.

### Exactly-Once and At-Most-Once Semantics

Using locks for tasks (as with the case of master election) is not sufficient to avoid having tasks executed multiple times because we can have, for example, the following succession of events:

1. Master *M1* assigns Task *T1* to Worker *W1*.
2. *W1* acquires the lock for *T1*, executes it, and releases the lock.
3. Master *M1* suspects that *W1* has crashed and reassigns Task *T1* to worker *W2*.
4. *W2* acquires the lock for *T1*, executes it, and releases the lock.

Here, the lock over *T1* did not prevent the task from being executed twice because the two workers did not interleave their steps when executing the task. To deal with cases in which exactly-once or at-most-once semantics are required, an application relies on mechanisms that are specific to its nature. For example, if application data has timestamps and a task is supposed to modify application data, then a successful execution of the task could be conditional on the timestamp values of the data it touches. The application also needs the ability to roll back partial changes in the case that the application state is not modified atomically; otherwise, it might end up with an inconsistent state.

The bottom line is that we are having this discussion just to illustrate the difficulties with implementing these kinds of semantics for applications. It is not within the scope of this book to discuss in detail the implementation of such semantics.

Another important issue with communication failures is the impact they have on synchronization primitives like locks. Because nodes can crash and systems are prone to network partitions, locks can be problematic: if a node crashes or gets partitioned away, the lock can prevent others from making progress. ZooKeeper consequently needs to implement mechanisms to deal with such scenarios. First, it enables clients to say that some data in the ZooKeeper state is *ephemeral*. Second, the ZooKeeper ensemble requires that clients periodically notify that they are alive. If a client fails to notify the ensemble in a timely manner, then all ephemeral state belonging to this client is deleted. Using these two mechanisms, we are able to prevent clients individually from bringing the application to a halt in the presence of crashes and communication failures.

Recall that we argued that in systems in which we cannot control the delay of messages it is not possible to tell if a client has crashed or if it is just slow. Consequently, when we suspect that a client has crashed, we actually need to react by assuming that it could just be slow, and that it may execute some other actions in the future.

## Summary of Tasks

From the preceding descriptions, we can extract the following requirements for our master-worker architecture:

*Master election*
: It is critical for progress to have a master available to assign tasks to workers.

*Crash detection*
: The master must be able to detect when workers crash or disconnect.

*Group membership management*
: The master must be able to figure out which workers are available to execute tasks.

*Metadata management*
: The master and the workers must be able to store assignments and execution statuses in a reliable manner.

Ideally, each of these tasks is exposed to the application in the form of a *primitive*, hiding completely the implementation details from the application developer. ZooKeeper provides key mechanisms to implement such primitives so that developers can implement the ones that best suit their needs and focus on the application logic. Throughout this book, we often refer to implementations of tasks like master election or crash detection as primitives because these are concrete tasks that distributed applications build upon.

# Why Is Distributed Coordination Hard?

Some of the complications of writing distributed applications are immediately apparent. For example, when our application starts up, somehow all of the different processes need to find the application configuration. Over time this configuration may change. We could shut everything down, redistribute configuration files, and restart, but that may incur extended periods of application downtime during reconfiguration.

Related to the configuration problem is the problem of group membership. As the load changes, we want to be able to add or remove new machines and processes.

The problems just described are functional problems that you can design solutions for as you implement your distributed application; you can test your solutions before deployment and be pretty sure that you have solved the problems correctly. The truly difficult problems you will encounter as you develop distributed applications have to do with *faults*—specifically, crashes and communication faults. These failures can crop up at any point, and it may be impossible to enumerate all the different corner cases that need to be handled.

**Byzantine Faults**

Byzantine faults are faults that may cause a component to behave in some arbitrary (and often unanticipated) way. Such a faulty component might, for example, corrupt application state or even behave maliciously. Systems that are built under the assumption that these faults can occur require a higher degree of replication and the use of security primitives. Although we acknowledge that there have been significant advances in the development of techniques to tolerate Byzantine faults in the academic literature, we haven't felt the need to adopt such techniques in ZooKeeper, and consequently we have avoided the additional complexity in the code base.

Failures also highlight a big difference between applications that run on a single machine and distributed applications: in distributed apps, partial failures can take place. When a single machine crashes, all the processes running on that machine fail. If there are multiple processes running on the machine and a process fails, the other processes can find out about the failure from the operating system. The operating system can also provide strong messaging guarantees between processes. All of this changes in a distributed environment: if a machine or process fails, other machines will keep running and may need to take over for the faulty processes. To handle faulty processes, the processes that are still running must be able to detect the failure; messages may be lost, and there may even be clock drift.

Ideally, we design our systems under the assumption that communication is asynchronous: the machines we use may experience clock drift and may experience communication failures. We make this assumption because these things do happen. Clocks drift all the time, we have all experienced occasional network problems, and unfortunately, failures also happen. What kinds of limits does this put on what we can do?

Well, let's take the simplest case. Let's assume that we have a distributed configuration that has been changing. This configuration is as simple as it can be: one bit. The processes in our application can start up once all running processes have agreed on the value of the configuration bit.

It turns out that a famous result in distributed computing, known as *FLP* after the authors Fischer, Lynch, and Patterson, proved that in a distributed system with asynchronous communication and process crashes, processes may not always agree on the one bit of configuration.[1] A similar result known as *CAP*, which stands for Consistency, Availability, and Partition-tolerance, says that when designing a distributed system we may want all three of those properties, but that no system can handle all three.[2] ZooKeeper has been designed with mostly consistency and availability in mind, although it also provides read-only capability in the presence of network partitions.

Okay, so we cannot have an ideal fault-tolerant, distributed, real-world system that transparently takes care of all problems that might ever occur. We can strive for a slightly less ambitious goal, though. First, we have to relax some of our assumptions and/or our goals. For example, we may assume that the clock is synchronized within some bounds; we may choose to be always consistent and sacrifice the ability to tolerate some network partitions; there may be times when a process may be running, but must act as if it is faulty because it cannot be sure of the state of the system. While these are compromises, they are compromises that have allowed us to build some rather impressive distributed systems.

## ZooKeeper Is a Success, with Caveats

Having pointed out that the perfect solution is impossible, we can repeat that ZooKeeper is not going to solve all the problems that the distributed application developer has to face. It does give the developer a nice framework to deal with these problems, though. There has been a lot of work over the years in distributed computing that ZooKeeper builds upon. Paxos[3] and virtual synchrony[4] have been particularly influential in the design of ZooKeeper. It deals with the changes and situations as they arise as seamlessly as possible, and gives developers a framework to deal with situations that arise that just cannot be handled automatically.

ZooKeeper was originally developed at Yahoo!, home to an abundance of large distributed applications. We noticed that the distributed coordination aspects of some applications were not treated appropriately, so systems were deployed with single points of failure or were brittle. On the other hand, other developers would spend so much

1. Michael J. Fischer, Nancy A. Lynch, and Michael S. Paterson. "Impossibility of Distributed Consensus with One Faulty Process." *Proceedings of the 2nd ACM SIGACT-SIGMOD Symposium on Principles of Database Systems*, (1983), doi:10.1145/588058.588060.

2. Seth Gilbert and Nancy Lynch. "Brewer's Conjecture and the Feasibility of Consistent, Available, Partition-Tolerant Web Services." *ACM SIGACT News*, 33:2 (2002), doi:10.1145/564585.564601.

3. Leslie Lamport. "The Part-Time Parliament." *ACM Transactions on Computer Systems*, 16:2 (1998): 133–169.

4. K. Birman and T. Joseph. "Exploiting Virtual Synchrony in Distributed Systems." *Proceedings of the 11th ACM Symposium on Operating Systems Principles*, (1987): 123–138.

time on the distributed coordination that they wouldn't have enough resources to focus on the application functionality. We also noticed that these applications all had some basic coordination requirements in common, so we set out to devise a general solution that contained some key elements that we could implement once and use in many different applications. ZooKeeper has proven to be far more general and popular than we had ever thought possible.

Over the years we have found that people can easily deploy a ZooKeeper cluster and develop applications for it—so easily, in fact, that some developers use it without completely understanding some of the cases that require the developer to make decisions that ZooKeeper cannot make by itself. One of the purposes of writing this book is to make sure that developers understand what they need to do to use ZooKeeper effectively and why they need to do it that way.

CHAPTER 2

# Getting to Grips with ZooKeeper

The previous chapter discussed the requirements of distributed applications at a high level and argued that they often have common requirements for coordination. We used the master-worker example, which is representative of a broad class of practical applications, to extract a few of the commonly used primitives we described there. We are now ready to present ZooKeeper, a service that enables the implementation of such primitives for coordination.

## ZooKeeper Basics

Several primitives used for coordination are commonly shared across many applications. Consequently, one way of designing a service used for coordination is to come up with a list of primitives, expose calls to create instances of each primitive, and manipulate these instances directly. For example, we could say that distributed locks constitute an important primitive and expose calls to create, acquire, and release locks.

Such a design, however, suffers from a couple of important shortcomings. First, we need to either come up with an exhaustive list of primitives used beforehand, or keep extending the API to introduce new primitives. Second, it does not give flexibility to the application using the service to implement primitives in the way that is most suitable for it.

We consequently have taken a different path with ZooKeeper. ZooKeeper does not expose primitives directly. Instead, it exposes a file system-like API comprised of a small set of calls that enables applications to implement their own primitives. We typically use *recipes* to denote these implementations of primitives. Recipes include ZooKeeper operations that manipulate small data nodes, called *znodes*, that are organized hierarchically as a tree, just like in a file system. Figure 2-1 illustrates a znode tree. The root node contains four more nodes, and three of those nodes have nodes under them. The leaf nodes are the data.

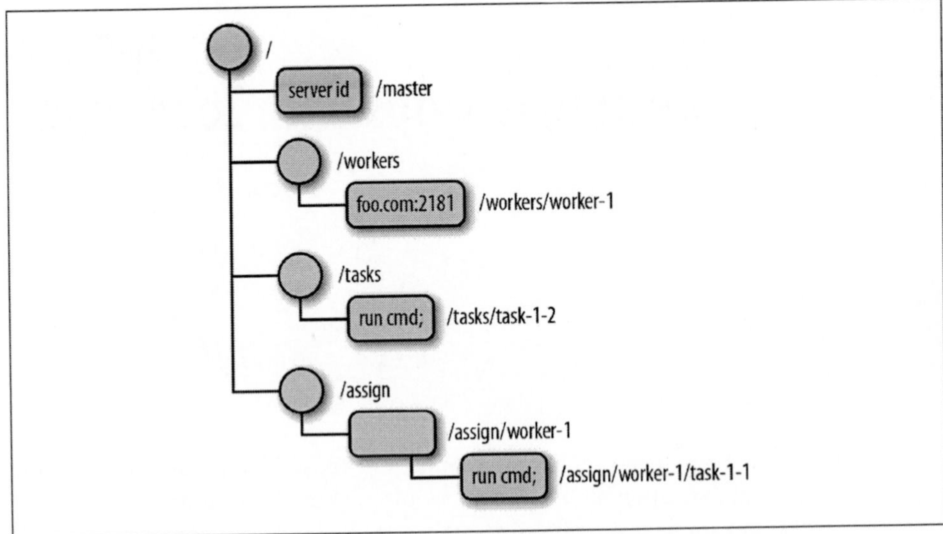


*Figure 2-1. ZooKeeper data tree example*

The absence of data often conveys important information about a znode. In a master-worker example, for instance, the absence of a master znode means that no master is currently elected. Figure 2-1 includes a few other znodes that could be useful in a master-worker configuration:

- The `/workers` znode is the parent znode to all znodes representing a worker available in the system. Figure 2-1 shows that one worker (`foo.com:2181`) is available. If a worker becomes unavailable, its znode should be removed from `/workers`.
- The `/tasks` znode is the parent of all tasks created and waiting for workers to execute them. Clients of the master-worker application add new znodes as children of `/tasks` to represent new tasks and wait for znodes representing the status of the task.
- The `/assign` znode is the parent of all znodes representing an assignment of a task to a worker. When a master assigns a task to a worker, it adds a child znode to `/assign`.

## API Overview

Znodes may or may not contain data. If a znode contains any data, the data is stored as a byte array. The exact format of the byte array is specific to each application, and ZooKeeper does not directly provide support to parse it. Serialization packages such as Protocol Buffers, Thrift, Avro, and MessagePack may be handy for dealing with the format of the data stored in znodes, but sometimes string encodings such as UTF-8 or ASCII suffice.

The ZooKeeper API exposes the following operations:

`create /path data`
: Creates a znode named with `/path` and containing `data`

`delete /path`
: Deletes the znode `/path`

`exists /path`
: Checks whether `/path` exists

`setData /path data`
: Sets the data of znode `/path` to `data`

`getData /path`
: Returns the data in `/path`

`getChildren /path`
: Returns the list of children under `/path`

One important note is that ZooKeeper does not allow partial writes or reads of the znode data. When setting the data of a znode or reading it, the content of the znode is replaced or read entirely.

ZooKeeper clients connect to a ZooKeeper service and establish a *session* through which they make API calls. If you are really anxious to use ZooKeeper, skip to "Sessions" on page 25. That section explains how to run some ZooKeeper commands from a command shell.

## Different Modes for Znodes

When creating a new znode, you also need to specify a *mode*. The different modes determine how the znode behaves.

### Persistent and ephemeral znodes

A znode can be either persistent or ephemeral. A *persistent* znode `/path` can be deleted only through a call to `delete`. An *ephemeral* znode, in contrast, is deleted if the client that created it crashes or simply closes its connection to ZooKeeper.

Persistent znodes are useful when the znode stores some data on behalf of an application and this data needs to be preserved even after its creator is no longer part of the system. For example, in the master-worker example, we need to maintain the assignment of tasks to workers even when the master that performed the assignment crashes.

Ephemeral znodes convey information about some aspect of the application that must exist only while the session of its creator is valid. For example, the master znode in our master-worker example is ephemeral. Its presence implies that there is a master and the master is up and running. If the master znode remains while the master is gone, then the system won't be able to detect the master crash. This would prevent the system from making progress, so the znode must go with the master. We also use ephemeral znodes for workers. If a worker becomes unavailable, its session expires and its znode in `/workers` disappears automatically.

An ephemeral znode can be deleted in two situations:

1. When the session of the client creator ends, either by expiration or because it explicitly closed.
2. When a client, not necessarily the creator, deletes it.

Because ephemeral znodes are deleted when their creator's session expires, we currently do not allow ephemerals to have children. There have been discussions in the community about allowing children for ephemeral znodes by making them also ephemeral. This feature might be available in future releases, but it isn't available currently.

### Sequential znodes

A znode can also be set to be *sequential*. A sequential znode is assigned a unique, monotonically increasing integer. This sequence number is appended to the path used to create the znode. For example, if a client creates a sequential znode with the path `/tasks/task-`, ZooKeeper assigns a sequence number, say 1, and appends it to the path. The path of the znode becomes `/tasks/task-1`. Sequential znodes provide an easy way to create znodes with unique names. They also provide a way to easily see the creation order of znodes.

To summarize, there are four options for the mode of a znode: `persistent`, `ephemeral`, `persistent_sequential`, and `ephemeral_sequential`.

## Watches and Notifications

Because ZooKeeper is typically accessed as a remote service, accessing a znode every time a client needs to know its content would be very expensive: it would induce higher latency and more operations to a ZooKeeper installation. Consider the example in Figure 2-2. The second call to `getChildren /tasks` returns the same value, an empty set, and consequently is unnecessary.

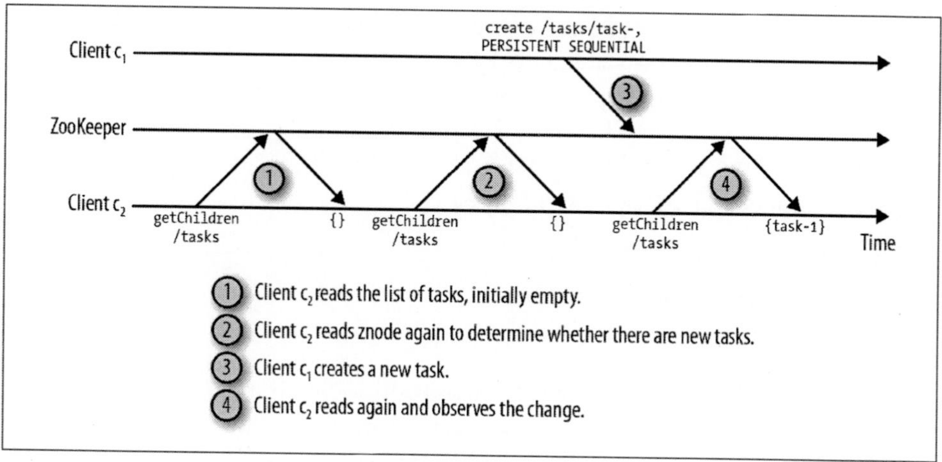


*Figure 2-2. Multiple reads to the same znode*

This is a common problem with polling. To replace the client polling, we have opted for a mechanism based on *notifications*: clients register with ZooKeeper to receive notifications of changes to znodes. Registering to receive a notification for a given znode consists of setting a *watch*. A watch is a one-shot operation, which means that it triggers one notification. To receive multiple notifications over time, the client must set a new watch upon receiving each notification. In the situation illustrated in Figure 2-3, the client reads new values from ZooKeeper only when it receives a notification indicating that the value of `/tasks` has changed.

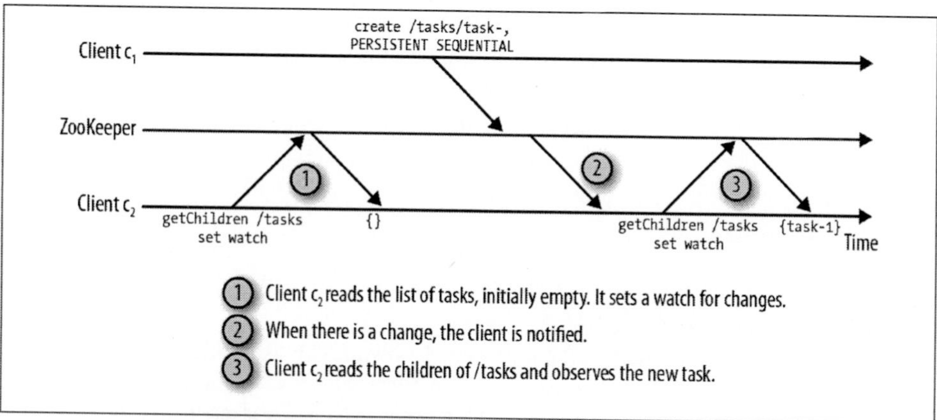


*Figure 2-3. Using notifications to be informed of changes to a znode*

When using notifications, there are a few things to be aware of. Because notifications are one-shot operations, it is possible that new changes will occur to a znode between a client receiving a notification for the znode and setting a new watch (but don't worry, you won't miss changes to the state). Let's look at a quick example to see how it works. Suppose the following events happen in this order:

1. Client $c_1$ sets a watch for changes to the data of `/tasks`.
2. Client $c_2$ comes and adds a task to `/tasks`.
3. Client $c_1$ receives the notification.
4. Client $c_1$ sets a new watch, but before it does it, a third client, $c_3$, comes and adds a new task to `/tasks`.

Client $c_1$ eventually has this new watch set, but no notification is triggered by the change made by $c_3$. To observe this change, $c_1$ has to actually read the state of `/tasks`, which it does when setting the watch because we set watches with operations that read the state of ZooKeeper. Consequently, $c_1$ does not miss any changes.

One important guarantee of notifications is that they are delivered to a client before any other change is made to the same znode. If a client sets a watch to a znode and there are two consecutive updates to the znode, the client receives the notification after the first update and before it has a chance to observe the second update by, say, reading the znode data. The key property we satisfy is the one that notifications preserve the order of updates the client observes. Although changes to the state of ZooKeeper may end up propagating more slowly to any given client, we guarantee that clients observe changes to the ZooKeeper state according to a global order.

ZooKeeper produces different types of notifications, depending on how the watch corresponding to the notification was set. A client can set a watch for changes to the data of a znode, changes to the children of a znode, or a znode being created or deleted. To set a watch, we can use any of the calls in the API that read the state of ZooKeeper. These API calls give the option of passing a `Watcher` object or using the default watcher. In our discussion of the master-worker example later in this chapter ("Implementation of a Master-Worker Example" on page 35) and in Chapter 4, we will cover how to use this mechanism in more detail.

**Who Manages My Cache?**

Instead of having the client manage its own cache of ZooKeeper values, we could have had ZooKeeper manage such a cache on behalf of the application. However, this would have made the design of ZooKeeper more complex. In fact, if ZooKeeper had to manage cache invalidations, it could cause ZooKeeper operations to stall while waiting for a client to acknowledge a cache invalidation request, because write operations would need a confirmation that all cached values had been invalidated.

## Versions

Each znode has a version number associated with it that is incremented every time its data changes. A couple of operations in the API can be executed conditionally: `setData` and `delete`. Both calls take a version as an input parameter, and the operation succeeds only if the version passed by the client matches the current version on the server. The use of versions is important when multiple ZooKeeper clients might be trying to perform operations over the same znode. For example, suppose that a client $c_1$ writes a znode `/config` containing some configuration. If another client $c_2$ concurrently updates the znode, the version $c_1$ has is stale and the `setData` of $c_1$ must not succeed. Using versions avoids such situations. In this case, the version that $c_1$ uses when writing back doesn't match and the operation fails. This situation is illustrated in Figure 2-4.

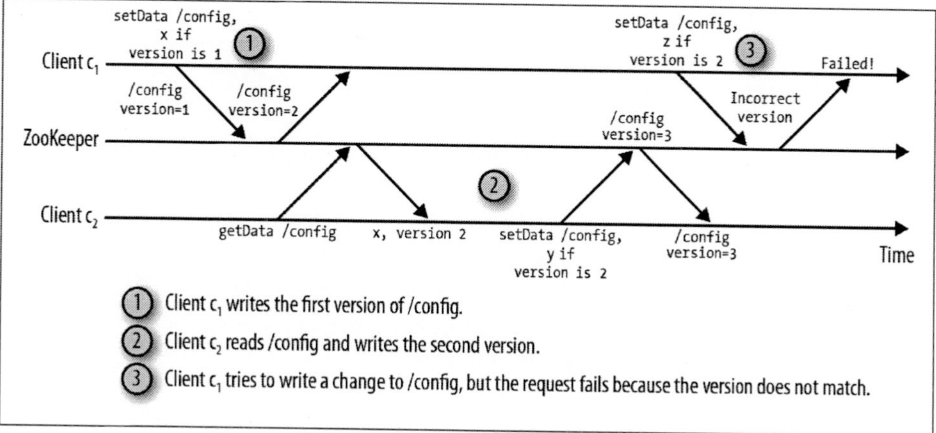


*Figure 2-4. Using versions to prevent inconsistencies due to concurrent updates*

# ZooKeeper Architecture

Now that we have discussed at a high level the operations that ZooKeeper exposes to applications, we need to understand more of how the service actually works. Applications make calls to ZooKeeper through a client library. The client library is responsible for the interaction with ZooKeeper servers.

Figure 2-5 shows the relationship between clients and servers. Each client imports the client library, and then can communicate with any ZooKeeper node.

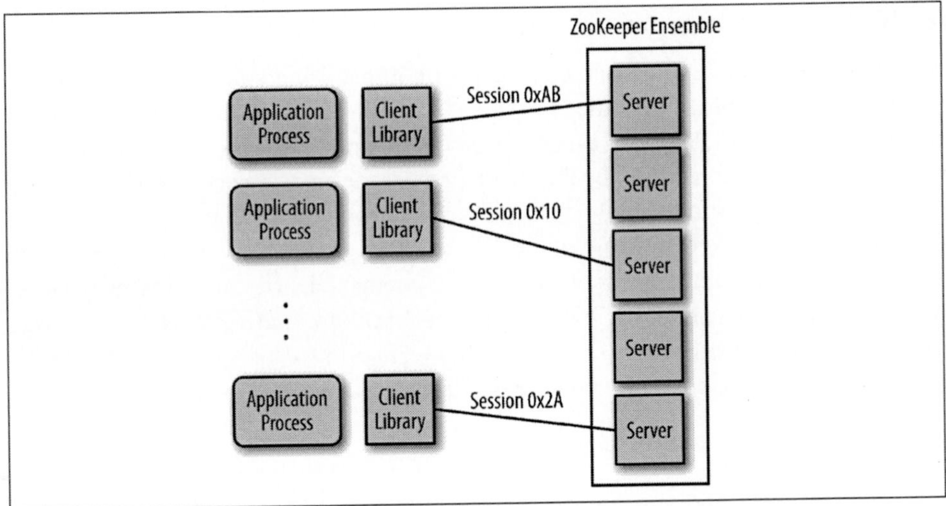


*Figure 2-5. ZooKeeper architecture overview*

ZooKeeper servers run in two modes: *standalone* and *quorum*. Standalone mode is pretty much what the term says: there is a single server, and ZooKeeper state is not replicated. In quorum mode, a group of ZooKeeper servers, which we call a *ZooKeeper ensemble*, replicates the state, and together they serve client requests. From this point on, we use the term "ZooKeeper ensemble" to denote an installation of servers. This installation could contain a single server and operate in standalone mode or contain a group of servers and operate in quorum mode.

## ZooKeeper Quorums

In quorum mode, ZooKeeper replicates its data tree across all servers in the ensemble. But if a client had to wait for every server to store its data before continuing, the delays might be unacceptable. In public administration, a *quorum* is the minimum number of legislators required to be present for a vote. In ZooKeeper, it is the minimum number of servers that have to be running and available in order for ZooKeeper to work. This number is also the minimum number of servers that have to store a client's data before telling the client it is safely stored. For instance, we might have five ZooKeeper servers in total, but a quorum of three. So long as any three servers have stored the data, the client can continue, and the other two servers will eventually catch up and store the data.

It is important to choose an adequate size for the quorum. Quorums must guarantee that, regardless of delays and crashes in the system, any update request the service positively acknowledges will persist until another request supersedes it.

To understand what this means, let's look at an example that shows how things can go wrong if the quorum is too small. Say we have five servers and a quorum can be any set of two servers. Now say that servers $s_1$ and $s_2$ acknowledge that they have replicated a request to create a znode /z. The service returns to the client saying that the znode has been created. Now suppose servers $s_1$ and $s_2$ are partitioned away from the other servers and from clients for an arbitrarily long time, before they have a chance to replicate the new znode to the other servers. The service in this state is able to make progress because there are three servers available and it really needs only two according to our assumptions, but these three servers have never seen the new znode /z. Consequently, the request to create /z is not durable.

This is an example of the split-brain scenario mentioned in Chapter 1. To avoid this problem, in this example the size of the quorum must be at least three, which is a majority out of the five servers in the ensemble. To make progress, the ensemble needs at least three servers available. To confirm that a request to update the state has completed successfully, this ensemble also requires that at least three servers acknowledge that they have replicated it. Consequently, if the ensemble is able to make progress, then for every update that has completed successfully, we have at least one server available that contains a copy of that update (that is, the possible quorums intersect in at least one node).

Using such a majority scheme, we are able to tolerate the crash of $f$ servers, where $f$ is less than half of the servers in the ensemble. For example, if we have five servers, we can tolerate up to $f = 2$ crashes. The number of servers in the ensemble is not mandatorily odd, but an even number actually makes the system more fragile. Say that we use four servers for an ensemble. A majority of servers is comprised of three servers. However, this system will only tolerate a single crash, because a double crash makes the system lose majority. Consequently, with four servers, we can only tolerate a single crash, but quorums now are larger, which implies that we need more acknowledgments for each request. The bottom line is that we should always shoot for an odd number of servers.

We allow quorum sets other than majority quorums, but that is a discussion for a more advanced chapter. We discuss it in Chapter 10.

## Sessions

Before executing any request against a ZooKeeper ensemble, a client must establish a session with the service. The concept of sessions is very important and quite critical for the operation of ZooKeeper. All operations a client submits to ZooKeeper are associated to a session. When a session ends for any reason, the ephemeral nodes created during that session disappear.

When a client creates a ZooKeeper handle using a specific language binding, it establishes a session with the service. The client initially connects to any server in the ensemble, and only to a single server. It uses a TCP connection to communicate with the server, but the session may be moved to a different server if the client has not heard from its current server for some time. Moving a session to a different server is handled transparently by the ZooKeeper client library.

Sessions offer *order guarantees*, which means that requests in a session are executed in FIFO (first in, first out) order. Typically, a client has only a single session open, so its requests are all executed in FIFO order. If a client has multiple concurrent sessions, FIFO ordering is not necessarily preserved across the sessions. Consecutive sessions of the same client, even if they don't overlap in time, also do not necessarily preserve FIFO order. Here is how it can happen in this case:

- Client establishes a session and makes two consecutive asynchronous calls to `create /tasks` and `/workers`.
- First session expires.
- Client establishes another session and makes an asynchronous call to `create /assign`.

In this sequence of calls, it is possible that only `/tasks` and `/assign` have been created, which preserves FIFO ordering for the first session but violates it across sessions.

## Getting Started with ZooKeeper

To get started, you need to download the ZooKeeper distribution. ZooKeeper is an Apache project hosted at *http://zookeeper.apache.org*. If you follow the download links, you should end up downloading a compressed TAR file named something like *zookeeper-3.4.5.tar.gz*. On Linux, Mac OS X, or any other UNIX-like system you can use the following command to extract the distribution:

```
# tar -xvzf zookeeper-3.4.5.tar.gz
```

If you are using Windows, you will need to use a decompression tool such as WinZip to extract the distribution.

You will also need to have Java installed. Java 6 is required to run ZooKeeper.

In the *distribution* directory you will find a *bin* directory that contains the scripts needed to start ZooKeeper. The scripts that end in *.sh* are designed to run on UNIX platforms (Linux, Mac OS X, etc.), and the scripts that end in *.cmd* are for Windows. The *conf* directory holds configuration files. The *lib* directory contains Java JAR files, which are third-party files needed to run ZooKeeper. Later we will need to refer to the directory in which you extracted the ZooKeeper distribution. We will refer to this directory as `{PATH_TO_ZK}`.

## First ZooKeeper Session

Let's set up ZooKeeper locally in standalone mode and create a session. To do this, we use the *zkServer* and *zkCli* tools that come with the ZooKeeper distribution under *bin/*. Experienced administrators typically use them for debugging and administration, but it turns out that they are great for getting beginners familiar with ZooKeeper as well.

Assuming you have downloaded and unpackaged a ZooKeeper distribution, go to a shell, change directory (cd) to the project's root, and rename the sample configuration file:

```
# mv conf/zoo_sample.cfg conf/zoo.cfg
```

Although optional, it might be a good idea to move the *data* directory out of */tmp* to avoid having ZooKeeper filling up your root partition. You can change its location in the *zoo.cfg* file:

```
dataDir=/users/me/zookeeper
```

To finally start a server, execute the following:

```
# bin/zkServer.sh start
JMX enabled by default
Using config: ../conf/zoo.cfg
Starting zookeeper ... STARTED
#
```

This server command makes the ZooKeeper server run in the background. To have it running in the foreground in order to see the output of the server, you can run the following:

```
# bin/zkServer.sh start-foreground
```

This option gives you much more verbose output and allows you to see what's going on with the server.

We are now ready to start a client. In a different shell under the project root, run the following:

```
# bin/zkCli.sh
.
.
.
<some omitted output>
.
.
.
2012-12-06 12:07:23,545 [myid:] - INFO  [main:ZooKeeper@438] -  ❶
Initiating client connection, connectString=localhost:2181
sessionTimeout=30000 watcher=org.apache.zookeeper.
ZooKeeperMain$MyWatcher@2c641e9a
Welcome to ZooKeeper!
2012-12-06 12:07:23,702 [myid:] - INFO  [main-SendThread ❷
```

```
(localhost:2181):ClientCnxn$SendThread@966] - Opening
socket connection to server localhost/127.0.0.1:2181.
Will not attempt to authenticate using SASL (Unable to
locate a login configuration)
JLine support is enabled
2012-12-06 12:07:23,717 [myid:] - INFO  [main-SendThread  ❸
(localhost:2181):ClientCnxn$SendThread@849] - Socket
connection established to localhost/127.0.0.1:2181, initiating
session [zk: localhost:2181(CONNECTING) 0]
2012-12-06 12:07:23,987 [myid:] - INFO  [main-SendThread  ❹
(localhost:2181):ClientCnxn$SendThread@1207] - Session
establishment complete on server localhost/127.0.0.1:2181,
sessionid = 0x13b6fe376cd0000, negotiated timeout = 30000

WATCHER::

WatchedEvent state:SyncConnected type:None path:null ❺
```

❶ The client starts the procedure to establish a session.

❷ The client tries to connect to `localhost/127.0.0.1:2181`.

❸ The client's connection is successful and the server proceeds to initiate a new session.

❹ The session initialization completes successfully.

❺ The server sends the client a `SyncConnected` event.

Let's have a look at the output. There are a number of lines just telling us how the various environment variables have been set and what JARs the client is using. We'll ignore them for the purposes of this example and focus on the session establishment, but take your time to analyze the full output on your screen.

Toward the end of the output, we see log messages referring to session establishment. The first one says "Initiating client connection." The message pretty much says what is happening, but an additional important detail is that it is trying to connect to one of the servers in the `localhost/127.0.0.1:2181` connect string sent by the client. In this case, the string contains only `localhost`, so this is the one the connection will go for. Next we see a message about SASL, which we will ignore, followed by a confirmation that the client successfully established a TCP connection with the local ZooKeeper server. The last message of the log confirms the session establishment and gives us its ID: `0x13b6fe376cd0000`. Finally, the client library notifies the application with a `SyncConnected` event. Applications are supposed to implement `Watcher` objects that process such events. We will talk more about events in the next section.

Just to get a bit more familiar with ZooKeeper, let's list the znodes under the root and create a znode. Let's first confirm that the tree is empty at this point, aside from the `/zookeeper` znode that marks the metadata tree that the ZooKeeper service maintains:

```
WATCHER::

WatchedEvent state:SyncConnected type:None path:null

[zk: localhost:2181(CONNECTED) 0] ls /
[zookeeper]
```

What happened here? We have executed `ls /` and we see that there is only `/zookeeper` there. Now we create a znode called `/workers` and make sure that it is there as follows:

```
WATCHER::

WatchedEvent state:SyncConnected type:None path:null

[zk: localhost:2181(CONNECTED) 0]
[zk: localhost:2181(CONNECTED) 0] ls /
[zookeeper]
[zk: localhost:2181(CONNECTED) 1] create /workers ""
Created /workers
[zk: localhost:2181(CONNECTED) 2] ls /
[workers, zookeeper]
[zk: localhost:2181(CONNECTED) 3]
```

### Znode Data

When we create the `/workers` znode, we specify an empty string (`""`) to say that there is no data we want to store in the znode at this point. That parameter in this interface, however, allows us to store arbitrary strings in the ZooKeeper znodes. For example, we could have replaced `""` with `"workers"`.

To wrap up this exercise, we delete the znode and exit:

```
[zk: localhost:2181(CONNECTED) 3] delete /workers
[zk: localhost:2181(CONNECTED) 4] ls /
[zookeeper]
[zk: localhost:2181(CONNECTED) 5] quit
Quitting...
2012-12-06 12:28:18,200 [myid:] - INFO  [main-EventThread:ClientCnxn$
EventThread@509] - EventThread shut down
2012-12-06 12:28:18,200 [myid:] - INFO  [main:ZooKeeper@684] - Session:
0x13b6fe376cd0000 closed
```

Observe that the znode `/workers` has been deleted and the session has now been closed. To clean up, let's also stop the ZooKeeper server:

```
# bin/zkServer.sh stop
JMX enabled by default
Using config: ../conf/zoo.cfg
Stopping zookeeper ... STOPPED
#
```

## States and the Lifetime of a Session

The *lifetime* of a session corresponds to the period between its creation and its end, whether it is closed gracefully or expires because of a timeout. To talk about what happens in a session, we need to consider the possible states that a session can be in and the possible events that can change its state.

The main possible states of a session are mostly self-explanatory: `CONNECTING`, `CONNECTED`, `CLOSED`, and `NOT_CONNECTED`. The state transitions depend on various events that happen between the client and the service (Figure 2-6).

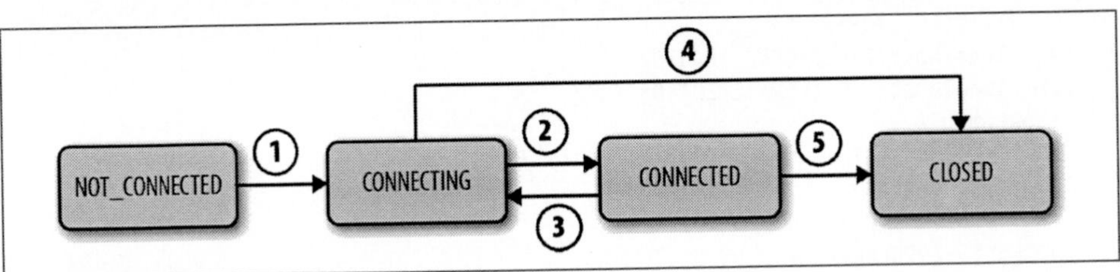


*Figure 2-6. Session states and transitions*

A session starts at the `NOT_CONNECTED` state and transitions to `CONNECTING` (arrow 1 in Figure 2-6) with the initialization of the ZooKeeper client. Normally, the connection to a ZooKeeper server succeeds and the session transitions to `CONNECTED` (arrow 2). When the client loses its connection to the ZooKeeper server or doesn't hear from the server, it transitions back to `CONNECTING` (arrow 3) and tries to find another ZooKeeper server. If it is able to find another server or to reconnect to the original server, it transitions back to `CONNECTED` once the server confirms that the session is still valid. Otherwise, it declares the session expired and transitions to `CLOSED` (arrow 4). The application can also explicitly close the session (arrows 4 and 5).

**Waiting on CONNECTING During Network Partitions**

If a client disconnects from a server due to a timeout, the client remains in the `CONNECTING` state. If the disconnection happens because the client has been partitioned away from the ZooKeeper ensemble, it will remain in this state until either it closes the session explicitly, or the partition heals and the client hears from a ZooKeeper server that the session has expired. We have this behavior because the ZooKeeper ensemble is the one responsible for declaring a session expired, not the client. Until the client hears that a ZooKeeper session has expired, the client cannot declare the session expired. The client may choose to close the session, however.

One important parameter you should set when creating a session is the session timeout, which is the amount of time the ZooKeeper service allows a session before declaring it expired. If the service does not see messages associated to a given session during time *t*, it declares the session expired. On the client side, if it has heard nothing from the server at 1/3 of *t*, it sends a heartbeat message to the server. At 2/3 of *t*, the ZooKeeper client starts looking for a different server, and it has another 1/3 of *t* to find one.

**Which Server Does It Try to Connect To?**

In quorum mode a client has multiple choices of servers to connect to, whereas in standalone mode it must try to reconnect to the single server available. In quorum mode, the application is supposed to pass a list of servers the client can connect to and choose from.

When trying to connect to a different server, it is important for the ZooKeeper state of this server to be at least as fresh as the last ZooKeeper state the client has observed. A client cannot connect to a server that has not seen an update that the client might have seen. ZooKeeper determines freshness by ordering updates in the service. Every change to the state of a ZooKeeper deployment is totally ordered with respect to all other executed updates. Consequently, if a client has observed an update in position *i*, it cannot connect to a server that has only seen $i' < i$. In the ZooKeeper implementation, the transaction identifiers that the system assigns to each update establish the order.

Figure 2-7 illustrates the use of transaction identifiers (zxids) for reconnecting. After the client disconnects from $s_1$ because it times out, it tries $s_2$, but $s_2$ has lagged behind and does not reflect a change known to the client. However, $s_3$ has seen the same changes as the client, so it is safe to connect.

## ZooKeeper with Quorums

The configuration we have used so far is for a standalone server. If the server is up, the service is up, but if the server fails, the whole service comes down with it. This doesn't quite live up to the promise of a reliable coordination service. To truly get reliability, we need to run multiple servers.

Fortunately, we can run multiple servers even if we only have a single machine. We just need to set up a more sophisticated configuration.

In order for servers to contact each other, they need some contact information. In theory servers can use multicasts to discover each other, but we want to support ZooKeeper ensembles that span multiple networks in addition to single networks that support multiple ensembles.

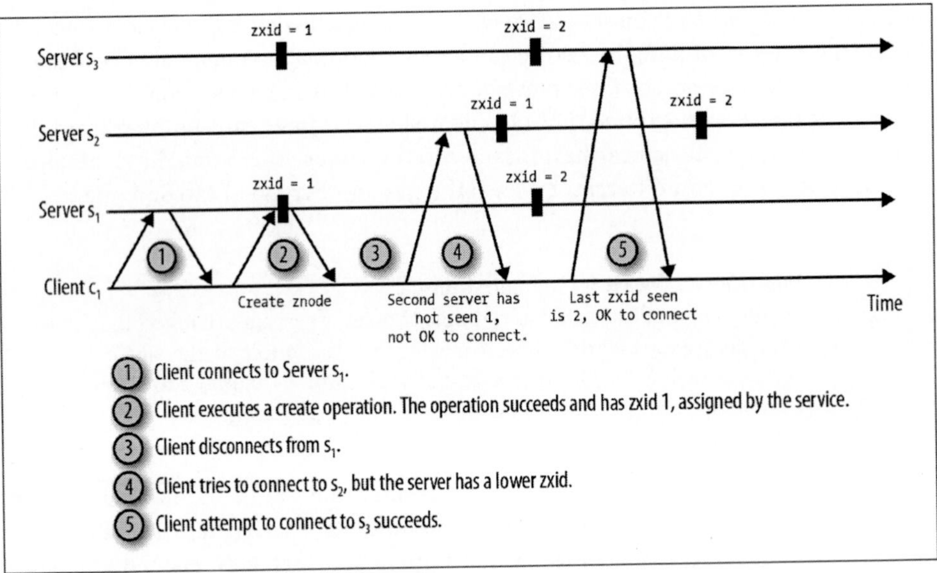


*Figure 2-7. Example of client reconnecting*

To accomplish this, we are going to use the following configuration file:

```
tickTime=2000
initLimit=10
syncLimit=5
dataDir=./data
clientPort=2181
server.1=127.0.0.1:2222:2223
server.2=127.0.0.1:3333:3334
server.3=127.0.0.1:4444:4445
```

We'll concentrate on the last three lines here, the `server.n` entries. The rest are general configuration parameters that will be explained in Chapter 10.

Each `server.`*`n`* entry specifies the address and port numbers used by ZooKeeper server *n*. There are three colon-separated fields for each `server.`*`n`* entry. The first field is the hostname or IP address of server *n*. The second and third fields are TCP port numbers used for quorum communication and leader election. Because we are starting up three server processes on the same machine, we need to use different port numbers for each entry. Normally, when running each server process on its own server, each server entry will use the same port numbers.

We also need to set up some *data* directories. We can do this from the command line with the following commands:

```
mkdir z1
mkdir z1/data
mkdir z2
```

```
mkdir z2/data
mkdir z3
mkdir z3/data
```

When we start up a server, it needs to know which server it is. A server figures out its ID by reading a file named *myid* in the *data* directory. We can create these files with the following commands:

```
echo 1 > z1/data/myid
echo 2 > z2/data/myid
echo 3 > z3/data/myid
```

When a server starts up, it finds the *data* directory using the `dataDir` parameter in the configuration file. It obtains the server ID from `mydata` and then uses the corresponding `server.`*n* entry to set up the ports it listens on. When we run ZooKeeper server processes on different machines, they can all use the same client ports and even use exactly the same configuration files. But for this example, running on one machine, we need to customize the client ports for each server.

So, first we create *z1/z1.cfg* using the configuration file we discussed at the beginning of this section. We then create configurations *z2/z2.cfg* and *z3/z3.cfg* by changing `clientPort` to be `2182` and `2183`, respectively.

Now we can start a server. Let's begin with *z1*:

```
$ cd z1
$ {PATH_TO_ZK}/bin/zkServer.sh start ./z1.cfg
```

Log entries for the server go to *zookeeper.out*. Because we have only started one of three ZooKeeper servers, the service will not be able to come up. In the log we should see entries of the form:

```
... [myid:1] - INFO  [QuorumPeer[myid=1]/...:2181:QuorumPeer@670] - LOOKING
... [myid:1] - INFO  [QuorumPeer[myid=1]/...:2181:FastLeaderElection@740] -
New election. My id =  1, proposed zxid=0x0
... [myid:1] - INFO  [WorkerReceiver[myid=1]:FastLeaderElection@542] -
Notification: 1 ..., LOOKING (my state)
... [myid:1] - WARN  [WorkerSender[myid=1]:QuorumCnxManager@368] - Cannot
open channel to 2 at election address /127.0.0.1:3334
       Java.net.ConnectException: Connection refused
           at java.net.PlainSocketImpl.socketConnect(Native Method)
           at java.net.PlainSocketImpl.doConnect(PlainSocketImpl.java:351)
```

The server is frantically trying to connect to other servers, and failing. If we start up another server, we can form a quorum:

```
$ cd z2
$ {PATH_TO_ZK}/bin/zkServer.sh start ./z2.cfg
```

If we examine the *zookeeper.out* log for the second server, we see:

```
... [myid:2] - INFO  [QuorumPeer[myid=2]/...:2182:Leader@345] - LEADING
- LEADER ELECTION TOOK - 279
... [myid:2] - INFO  [QuorumPeer[myid=2]/...:2182:FileTxnSnapLog@240] -
Snapshotting: 0x0 to ./data/version-2/snapshot.0
```

This indicates that server 2 has been elected leader. If we now look at the log for server 1, we see:

```
... [myid:1] - INFO  [QuorumPeer[myid=1]/...:2181:QuorumPeer@738] -
FOLLOWING
... [myid:1] - INFO  [QuorumPeer[myid=1]/...:2181:ZooKeeperServer@162] -
Created server ...
... [myid:1] - INFO  [QuorumPeer[myid=1]/...:2181:Follower@63] - FOLLOWING
- LEADER ELECTION TOOK - 212
```

Server 1 has also become active as a follower of server 2. We now have a quorum of servers (two out of three) available.

At this point the service is available. Now we need to configure the clients to connect to the service. The connect string lists all the ***host:port*** pairs of the servers that make up the service. In this case, that string is `"127.0.0.1:2181, 127.0.0.1:2182, 127.0.0.1:2183"`. (We're including the third server, even though we never started it, because it can illustrate some useful properties of ZooKeeper.)

We use *zkCli.sh* to access the cluster using:

```
$ {PATH_TO_ZK}/bin/zkCli.sh -server 127.0.0.1:2181,127.0.0.1:2182,127.0.0.1:2183
```

When the server connects, we should see a message of the form:

```
[myid:] - INFO  [...] - Session establishment
complete on server localhost/127.0.0.1:2182 ...
```

Note the port number, 2182 in this case, in the log message. If we stop the client with Ctrl-C and restart it various times, we will see the port number change between 2181 and 2182. We may also notice a failed connection attempt to 2183 followed by a successful connection attempt to one of the other ports.

**Simple Load Balancing**

Clients connect in a random order to servers in the connect string. This allows ZooKeeper to achieve simple load balancing. However, it doesn't allow clients to specify a preference for a server to connect to. For example, if we have an ensemble of five ZooKeeper servers with three on the West Coast and two on the East Coast, to make sure clients connect only to local servers we may want to put only the East Coast servers in the connect string of the East Coast clients, and only the West Coast servers in the connect string of the West Coast clients.

The connection attempts to show how you can achieve reliability by running multiple servers (but in a production environment, of course, you will do so on different physical hosts). For most of this book, including the next few chapters, we stick with a standalone server for development because it is much easier to start and manage and makes the examples more straightforward. One of the nice things about ZooKeeper is that, apart from the connect string, it doesn't matter to the clients how many servers make up the ZooKeeper service.

## Implementing a Primitive: Locks with ZooKeeper

One simple example of what we can do with ZooKeeper is implement critical sections through locks. There are multiple flavors of locks (e.g., read/write locks, global locks) and several ways to implement locks with ZooKeeper. Here we discuss a simple recipe just to illustrate how applications can use ZooKeeper; we do not consider other variants of locks.

Say that we have an application with *n* processes trying to acquire a lock. Recall that ZooKeeper does not expose primitives directly, so we need to use the ZooKeeper interface to manipulate znodes and implement the lock. To acquire a lock, each process *p* tries to create a znode, say `/lock`. If *p* succeeds in creating the znode, it has the lock and can proceed to execute its critical section. One potential problem is that *p* could crash and never release the lock. In this case, no other process will ever be able to acquire the lock again, and the system could seize up in a deadlock. To avoid such situations, we just have to make the `/lock` znode ephemeral when we create it.

Other processes that try to create `/lock` fail so long as the znode exists. So, they watch for changes to `/lock` and try to acquire the lock again once they detect that `/lock` has been deleted. Upon receiving a notification that `/lock` has been deleted, if a process *p'* is still interested in acquiring the lock, it repeats the steps of attempting to create `/lock` and, if another process has created the znode already, watching it.

# Implementation of a Master-Worker Example

In this section we will implement some of the functionality of the master-worker example using the *zkCli* tool. This example is for didactic purposes only, and we do not recommend building systems using *zkCli*. The goal of using *zkCli* is simply to illustrate how we implement coordination recipes with ZooKeeper, putting aside much of the detail needed for a real implementation. We'll get into implementation details in the next chapter.

The master-worker model involves three roles:

*Master*
: The master watches for new workers and tasks, assigning tasks to available workers.

*Worker*
: Workers register themselves with the system, to make sure that the master sees they are available to execute tasks, and then watch for new tasks.

*Client*
: Clients create new tasks and wait for responses from the system.

Let's now go over the different roles and the exact steps each role needs to perform.

## The Master Role

Because only one process should be a master, a process must lock mastership in ZooKeeper to become the master. To do this, the process creates an ephemeral znode called `/master`:

```
[zk: localhost:2181(CONNECTED) 0] create -e /master "master1.example.com:2223" ❶
Created /master
[zk: localhost:2181(CONNECTED) 1] ls / ❷
[master, zookeeper]
[zk: localhost:2181(CONNECTED) 2] get /master ❸
"master1.example.com:2223"
cZxid = 0x67
ctime = Tue Dec 11 10:06:19 CET 2012
mZxid = 0x67
mtime = Tue Dec 11 10:06:19 CET 2012
pZxid = 0x67
cversion = 0
dataVersion = 0
aclVersion = 0
ephemeralOwner = 0x13b891d4c9e0005
dataLength = 26
numChildren = 0
[zk: localhost:2181(CONNECTED) 3]
```

❶ Create a master znode to get mastership. We use the `-e` flag to indicate that we are creating an ephemeral znode.

❷ List the root of the ZooKeeper tree.

❸ Get the metadata and data of the `/master` znode.

What has just happened? We first created an ephemeral znode with path `/master`. We added the host information to the znode in case others need to contact it outside ZooKeeper. It is not strictly necessary to add the host information, but we did so just to show that we can add data if needed. To make the znode ephemeral, we added the `-e` flag. Remember that an ephemeral node is automatically deleted if the session in which it has been created closes or expires.

Let's say now that we use two processes for the master role, although at any point in time there can be at most one active master. The other process is a backup master. Say that the other process, unaware that a master has been elected already, also tries to create a /master znode. Let's see what happens:

```
[zk: localhost:2181(CONNECTED) 0] create -e /master "master2.example.com:2223"
Node already exists: /master
[zk: localhost:2181(CONNECTED) 1]
```

ZooKeeper tells us that a /master node already exists. This way, the second process knows that there is already a master. However, it is possible that the active master may crash, and the backup master may need to take over the role of active master. To detect this, we need to set a watch on the /master node as follows:

```
[zk: localhost:2181(CONNECTED) 0] create -e /master "master2.example.com:2223"
Node already exists: /master
[zk: localhost:2181(CONNECTED) 1] stat /master true
cZxid = 0x67
ctime = Tue Dec 11 10:06:19 CET 2012
mZxid = 0x67
mtime = Tue Dec 11 10:06:19 CET 2012
pZxid = 0x67
cversion = 0
dataVersion = 0
aclVersion = 0
ephemeralOwner = 0x13b891d4c9e0005
dataLength = 26
numChildren = 0
[zk: localhost:2181(CONNECTED) 2]
```

The stat command gets the attributes of a znode and allows us to set a watch on the existence of the znode. Having the parameter true after the path sets the watch. In the case that the active primary crashes, we observe the following:

```
[zk: localhost:2181(CONNECTED) 0] create -e /master "master2.example.com:2223"
Node already exists: /master
[zk: localhost:2181(CONNECTED) 1] stat /master true
cZxid = 0x67
ctime = Tue Dec 11 10:06:19 CET 2012
mZxid = 0x67
mtime = Tue Dec 11 10:06:19 CET 2012
pZxid = 0x67
cversion = 0
dataVersion = 0
aclVersion = 0
ephemeralOwner = 0x13b891d4c9e0005
dataLength = 26
numChildren = 0
[zk: localhost:2181(CONNECTED) 2]
WATCHER::

WatchedEvent state:SyncConnected type:NodeDeleted path:/master

[zk: localhost:2181(CONNECTED) 2] ls /
```

```
[zookeeper]
[zk: localhost:2181(CONNECTED) 3]
```

Note the `NodeDeleted` event at the end of the output. This event indicates that the active primary has had its session closed, or it has expired. Note also that the `/master` znode is no longer there. The backup primary should now try to become the active primary by trying to create the `/master` znode again:

```
[zk: localhost:2181(CONNECTED) 0] create -e /master "master2.example.com:2223"
Node already exists: /master
[zk: localhost:2181(CONNECTED) 1] stat /master true
cZxid = 0x67
ctime = Tue Dec 11 10:06:19 CET 2012
mZxid = 0x67
mtime = Tue Dec 11 10:06:19 CET 2012
pZxid = 0x67
cversion = 0
dataVersion = 0
aclVersion = 0
ephemeralOwner = 0x13b891d4c9e0005
dataLength = 26
numChildren = 0
[zk: localhost:2181(CONNECTED) 2]
WATCHER::

WatchedEvent state:SyncConnected type:NodeDeleted path:/master

[zk: localhost:2181(CONNECTED) 2] ls /
[zookeeper]
[zk: localhost:2181(CONNECTED) 3] create -e /master "master2.example.com:2223"
Created /master
[zk: localhost:2181(CONNECTED) 4]
```

Because it succeeds in creating the `/master` znode, the client now becomes the active master.

## Workers, Tasks, and Assignments

Before we discuss the steps taken by workers and clients, let's first create three important parent znodes, `/workers`, `/tasks`, and `/assign`:

```
[zk: localhost:2181(CONNECTED) 0] create /workers ""
Created /workers
[zk: localhost:2181(CONNECTED) 1] create /tasks ""
Created /tasks
[zk: localhost:2181(CONNECTED) 2] create /assign ""
Created /assign
[zk: localhost:2181(CONNECTED) 3] ls /
[assign, tasks, workers, master, zookeeper]
[zk: localhost:2181(CONNECTED) 4]
```

The three new znodes are persistent znodes and contain no data. We use these znodes in this example to tell us which workers are available, tell us when there are tasks to assign, and make assignments to workers.

In a real application, these znodes need to be created either by a primary process before it starts assigning tasks or by some bootstrap procedure. Regardless of how they are created, once they exist, the master needs to watch for changes in the children of `/workers` and `/tasks`:

```
[zk: localhost:2181(CONNECTED) 4] ls /workers true
[]
[zk: localhost:2181(CONNECTED) 5] ls /tasks true
[]
[zk: localhost:2181(CONNECTED) 6]
```

Note that we have used the optional `true` parameter with `ls`, as we did before with `stat` on the master. The `true` parameter, in this case, creates a watch for changes to the set of children of the corresponding znode.

## The Worker Role

The first step by a worker is to notify the master that it is available to execute tasks. It does so by creating an ephemeral znode representing it under `/workers`. Workers use their hostnames to identify themselves:

```
[zk: localhost:2181(CONNECTED) 0] create -e /workers/worker1.example.com
                                             "worker1.example.com:2224"
Created /workers/worker1.example.com
[zk: localhost:2181(CONNECTED) 1]
```

Note from the output that ZooKeeper confirms that the znode has been created. Recall that the master has set a watch for changes to the children of `/workers`. Once the worker creates a znode under `/workers`, the master observes the following notification:

```
WATCHER::

WatchedEvent state:SyncConnected type:NodeChildrenChanged path:/workers
```

Next, the worker needs to create a parent znode, `/assign/worker1.example.com`, to receive assignments, and watch it for new tasks by executing `ls` with the second parameter set to `true`:

```
[zk: localhost:2181(CONNECTED) 0] create -e /workers/worker1.example.com
                                             "worker1.example.com:2224"
Created /workers/worker1.example.com
[zk: localhost:2181(CONNECTED) 1] create /assign/worker1.example.com ""
Created /assign/worker1.example.com
[zk: localhost:2181(CONNECTED) 2] ls /assign/worker1.example.com true
[]
[zk: localhost:2181(CONNECTED) 3]
```

The worker is now ready to receive assignments. We will look at assignments next, as we discuss the role of clients.

## The Client Role

Clients add tasks to the system. For the purposes of this example, it doesn't matter what the task really consists of. Here we assume that the client asks the master-worker system to run a command `cmd`. To add a task to the system, a client executes the following:

```
[zk: localhost:2181(CONNECTED) 0] create -s /tasks/task- "cmd"
Created /tasks/task-0000000000
```

We make the task znode sequential to create an order for the tasks added, essentially providing a queue. The client now has to wait until the task is executed. The worker that executes the task creates a status znode for the task once the task completes. The client determines that the task has been executed when it sees that a status znode for the task has been created; the client consequently must watch for the creation of the status znode:

```
[zk: localhost:2181(CONNECTED) 1] ls /tasks/task-0000000000 true
[]
[zk: localhost:2181(CONNECTED) 2]
```

The worker that executes the task creates a status znode as a child of `/tasks/task-0000000000`. That is the reason for watching the children of `/tasks/task-0000000000` with `ls`.

Once the task znode is created, the master observes the following event:

```
[zk: localhost:2181(CONNECTED) 6]
WATCHER::

WatchedEvent state:SyncConnected type:NodeChildrenChanged path:/tasks
```

The master next checks the new task, gets the list of available workers, and assigns it to `worker1.example.com`:

```
[zk: 6] ls /tasks
[task-0000000000]
[zk: 7] ls /workers
[worker1.example.com]
[zk: 8] create /assign/worker1.example.com/task-0000000000 ""
Created /assign/worker1.example.com/task-0000000000
[zk: 9]
```

The worker receives a notification that a new task has been assigned:

```
[zk: localhost:2181(CONNECTED) 3]
WATCHER::

WatchedEvent state:SyncConnected type:NodeChildrenChanged
path:/assign/worker1.example.com
```

The worker then checks the new task and sees that the task has been assigned to it:

```
WATCHER::

WatchedEvent state:SyncConnected type:NodeChildrenChanged
path:/assign/worker1.example.com

[zk: localhost:2181(CONNECTED) 3] ls /assign/worker1.example.com
[task-0000000000]
[zk: localhost:2181(CONNECTED) 4]
```

Once the worker finishes executing the task, it adds a status znode to `/tasks`:

```
[zk: localhost:2181(CONNECTED) 4] create /tasks/task-0000000000/status "done"
Created /tasks/task-0000000000/status
[zk: localhost:2181(CONNECTED) 5]
```

and the client receives a notification and checks the result:

```
WATCHER::

WatchedEvent state:SyncConnected type:NodeChildrenChanged
path:/tasks/task-0000000000

[zk: localhost:2181(CONNECTED) 2] get /tasks/task-0000000000
"cmd"
cZxid = 0x7c
ctime = Tue Dec 11 10:30:18 CET 2012
mZxid = 0x7c
mtime = Tue Dec 11 10:30:18 CET 2012
pZxid = 0x7e
cversion = 1
dataVersion = 0
aclVersion = 0
ephemeralOwner = 0x0
dataLength = 5
numChildren = 1
[zk: localhost:2181(CONNECTED) 3] get /tasks/task-0000000000/status
"done"
cZxid = 0x7e
ctime = Tue Dec 11 10:42:41 CET 2012
mZxid = 0x7e
mtime = Tue Dec 11 10:42:41 CET 2012
pZxid = 0x7e
cversion = 0
dataVersion = 0
aclVersion = 0
ephemeralOwner = 0x0
dataLength = 8
numChildren = 0
[zk: localhost:2181(CONNECTED) 4]
```

The client checks the content of the status znode to determine what has happened to the task. In this case, it has been successfully executed and the result is "done." Tasks can of course be more sophisticated and even involve another distributed system. The bottom line here is that regardless of what the task actually is, the mechanism to execute it and convey the results through ZooKeeper is in essence the same.

## Takeaway Messages

We have gone through a number of basic ZooKeeper concepts in this chapter. We have seen the basic functionality that ZooKeeper offers through its API and explored some important concepts about its architecture, like the use of quorums for replication. It is not important at this point to understand how the ZooKeeper replication protocol works, but it is important to understand the notion of quorums because you specify the number of servers when deploying ZooKeeper. Another important concept discussed here is sessions. Session semantics are critical for the ZooKeeper guarantees because they mostly refer to sessions.

To provide a preliminary understanding of how to work with ZooKeeper, we have used the *zkCli* tool to access a ZooKeeper server and execute requests against it. We have shown the main operations of the master-worker example executed with this tool. When implementing a real ZooKeeper application, you should not use this tool; it is there mostly for debugging and monitoring purposes. Instead, you'll use one of the language bindings ZooKeeper offers. In the next chapters, we will be using Java to implement our examples.

# PART II
# Programming with ZooKeeper

This part of the book can be read by programmers to develop skills and the right approaches to using ZooKeeper for coordination in their distributed programs. ZooKeeper comes with API bindings for Java and C. Both bindings have the same basic structures and signatures. Because the Java binding is the most popular and easiest to use, we will be using this binding in our examples. Chapter 7 introduces the C binding. The source code for the master-worker example is available in a GitHub repository.

CHAPTER 3

# Getting Started with the ZooKeeper API

In the previous chapter we used *zkCli* to introduce the basic ZooKeeper operations. In this chapter we are going to see how we actually use the API in applications. Here we give an introduction of how to program with the ZooKeeper API, showing how to create a session and implement a watcher. We also start coding our master-worker example.

## Setting the ZooKeeper CLASSPATH

We need to set up the appropriate classpath to run and compile ZooKeeper Java code. ZooKeeper uses a number of third-party libraries in addition to the ZooKeeper JAR file. To make typing a little easier and to make the text a little more readable we will use an environment variable `CLASSPATH` with all the required libraries. The script *zkEnv.sh* in the *bin* directory of the ZooKeeper distribution sets this environment variable for us. We need to source it using the following:

```
ZOOBINDIR="<path_to_distro>/bin"
. "$ZOOBINDIR"/zkEnv.sh
```

(On Windows, use the `call` command instead of the period and the *zkEnv.cmd* script.)

Once we run this script, the `CLASSPATH` variable will be set correctly. We will use it to compile and run our Java programs.

## Creating a ZooKeeper Session

The ZooKeeper API is built around a ZooKeeper *handle* that is passed to every API call. This handle represents a session with ZooKeeper. As shown in Figure 3-1, a session that is established with one ZooKeeper server will migrate to another ZooKeeper server if its connection is broken. As long as the session is alive, the handle will remain valid, and the ZooKeeper client library will continually try to keep an active connection to a ZooKeeper server to keep the session alive. If the handle is closed, the ZooKeeper client library will tell the ZooKeeper servers to kill the session. If ZooKeeper decides that a client has died, it will invalidate the session. If a client later tries to reconnect to a ZooKeeper server using the handle that corresponds to the invalidated session, the ZooKeeper server informs the client library that the session is no longer valid and the handle returns errors for all operations.

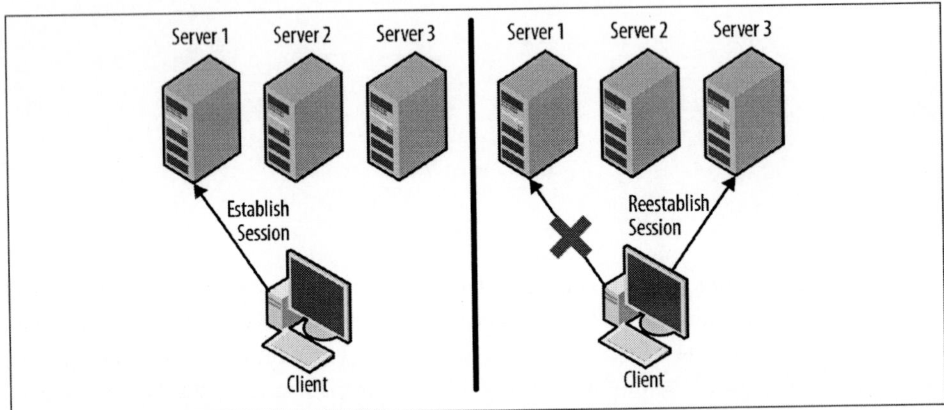


*Figure 3-1. Session migration between two servers*

The constructor that creates a ZooKeeper handle usually looks like:

```
ZooKeeper(
    String connectString,
    int sessionTimeout,
    Watcher watcher)
```

where:

`connectString`
: Contains the hostnames and ports of the ZooKeeper servers. We listed those servers when we used *zkCli* to connect to the ZooKeeper service.

`sessionTimeout`
: The time in milliseconds that ZooKeeper waits without hearing from the client before declaring the session dead. For now we will just use a value of `15000`, for 15 seconds. This means that if ZooKeeper cannot communicate with a client for more than 15 seconds, ZooKeeper will terminate the client's session. Note that this timeout is rather high, but it is useful for the experiments we will be doing later. ZooKeeper sessions typically have a timeout of 5–10 seconds.

`watcher`
: An object we need to create that will receive session events. Because `Watcher` is an interface, we will need to implement a class and then instantiate it to pass an instance to the ZooKeeper constructor. Clients use the `Watcher` interface to monitor the

health of the session with ZooKeeper. Events will be generated when a connection is established or lost to a ZooKeeper server. They can also be used to monitor changes to ZooKeeper data. Finally, if a session with ZooKeeper expires, an event is delivered through the `Watcher` interface to notify the client application.

## Implementing a Watcher

To receive notifications from ZooKeeper, we need to implement watchers. Let's look a bit more closely at the `Watcher` interface. It has the following declaration:

```
public interface Watcher {
    void process(WatchedEvent event);
}
```

Not much of an interface, right? We'll be using it heavily, but for now we will simply print the event. So, let's start our example implementation of the `Master`:

```
import org.apache.zookeeper.ZooKeeper;
import org.apache.zookeeper.Watcher;

public class Master implements Watcher {
    ZooKeeper zk;
    String hostPort;

    Master(String hostPort) {
        this.hostPort = hostPort; ❶
    }

    void startZK() {
        zk = new ZooKeeper(hostPort, 15000, this); ❷
    }

    public void process(WatchedEvent e) {
        System.out.println(e); ❸
    }

    public static void main(String args[])
        throws Exception {
        Master m = new Master(args[0]);

        m.startZK();

        // wait for a bit
        Thread.sleep(60000); ❹
    }
}
```

❶ In the constructor, we do not actually instantiate a `ZooKeeper` object. Instead, we just save the `hostPort` for later. Java best practices dictate that other methods of an object should not be called until the object's constructor has finished. Because this object implements `Watcher` and because once we instantiate a `ZooKeeper` object its `Watcher` callback may be invoked, we must construct the `ZooKeeper` object after the `Master` constructor has returned.

❷ Construct the `ZooKeeper` object using the `Master` object as the `Watcher` callback.

❸ This simple example does not have complex event handling. Instead, we will simply print out the event that we receive.

❹ Once we have connected to ZooKeeper, there will be a background thread that will maintain the ZooKeeper session. This thread is a daemon thread, which means that the program may exit even if the thread is still active. Here we sleep for a bit so that we can see some events come in before the program exits.

We can compile this simple example using the following:

```
$ javac  -cp $CLASSPATH Master.java
```

Once we have compiled *Master.java*, we run it and see the following:

```
$ java  -cp $CLASSPATH Master 127.0.0.1:2181
... - INFO  [...] - Client environment:zookeeper.version=3.4.5-1392090, ... ❶
...
... - INFO  [...] - Initiating client connection,
connectString=127.0.0.1:2181 ... ❷
... - INFO  [...] - Opening socket connection to server
localhost/127.0.0.1:2181. ...
... - INFO  [...] - Socket connection established to localhost/127.0.0.1:2181,
initiating session
... - INFO  [...] - Session establishment complete on server
localhost/127.0.0.1:2181, ... ❸
WatchedEvent state:SyncConnected type:None path:null ❹
```

The ZooKeeper client API produces various log messages to help the user understand what is happening. The logging is rather verbose and can be disabled using configuration files, but these messages are invaluable during development and even more invaluable if something unexpected happens after deployment.

❶ The first few log messages describe the ZooKeeper client implementation and environment.

❷ These log messages will be generated whenever a client initiates a connection to a ZooKeeper server. This may be the initial connection or subsequent reconnections.

❸ This message shows information about the connection after it has been established. It shows the host and port that the client connected to and the actual session timeout that was negotiated. If the requested session timeout is too short to be detected by the server or too long, the server will adjust the session timeout.

❹ This final line did not come from the ZooKeeper library; it is the `WatchEvent` object that we print in our implementation of `Watcher.process(WatchedEvent e)`.

This example run assumes that all the needed libraries are in the *lib* subdirectory. It also assumes that the configuration of *log4j.conf* is in the *conf* subdirectory. You can find these two directories in the ZooKeeper distribution that you are using. If you see the following:

```
log4j:WARN No appenders could be found for logger
      (org.apache.zookeeper.ZooKeeper).
log4j:WARN Please initialize the log4j system properly.
```

it means that you have not put the *log4j.conf* file in the classpath.

## Running the Watcher Example

What would have happened if we had started the master without starting the ZooKeeper service? Give it a try. Stop the service, then run `Master`. What do you see? The last line in the previous output, with the `WatchedEvent`, is not present. The ZooKeeper library isn't able to connect to the ZooKeeper server, so it doesn't tell us anything.

Now try starting the server, starting the `Master`, and then stopping the server while the `Master` is still running. You should see the `SyncConnected` event followed by the `Disconnected` event.

When developers see the `Disconnected` event, some think they need to create a new ZooKeeper handle to reconnect to the service. Do not do that! See what happens when you start the server, start the `Master`, and then stop and start the server while the `Master` is still running. You should see the `SyncConnected` event followed by the `Disconnected` event and then another `SyncConnected` event. The ZooKeeper client library takes care of reconnecting to the service for you. Unfortunately, network outages and server failures happen. Usually, ZooKeeper can deal with these failures.

It is important to remember that these failures also happen to ZooKeeper itself, however. A ZooKeeper server may fail or lose network connectivity, which will cause the scenario we simulated when we stopped the master. As long as the ZooKeeper service is made up of at least three servers, the failure of a single server will not cause a service outage. Instead, the client will see a `Disconnected` event quickly followed by a `SyncConnected` event.

**ZooKeeper Manages Connections**

Don't try to manage ZooKeeper client connections yourself. The ZooKeeper client library monitors the connection to the service and not only tells you about connection problems, but actively tries to reestablish communication. Often the library can quickly reestablish the session with minimal disruption to your application. Needlessly closing and starting a new session just causes more load on the system and longer availability outages.

It doesn't look like the client is doing much besides sleeping, but as we have seen by the events coming in, there are some things happening behind the scenes. We can also see what is happening on the ZooKeeper service side. ZooKeeper has two main management interfaces: JMX and four-letter words. We discuss these interfaces in depth in Chapter 10. Right now we will use the `stat` and `dump` four-letter words to see what is happening on the server.

To use these words, we need to *telnet* to the client port, 2181, and type them in (pressing the Enter key after the word). For example, if we start the `Master` and use the `stat` four-letter word, we should see the following:

```
$ telnet 127.0.0.1 2181
Trying 127.0.0.1...
Connected to 127.0.0.1.
Escape character is '^]'.
stat
ZooKeeper version: 3.4.5-1392090, built on 09/30/2012 17:52 GMT
Clients:
 /127.0.0.1:39470[1](queued=0,recved=3,sent=3)
 /127.0.0.1:39471[0](queued=0,recved=1,sent=0)

Latency min/avg/max: 0/5/48
Received: 34
Sent: 33
Connections: 2
Outstanding: 0
Zxid: 0x17
Mode: standalone
Node count: 4
Connection closed by foreign host.
```

We see from this output that there are two clients connected to the ZooKeeper server. One is `Master`, and the other is the Telnet connection.

If we start the `Master` and use the `dump` four-letter word, we should see the following:

```
$ telnet 127.0.0.1 2181
Trying 127.0.0.1...
Connected to 127.0.0.1.
```

```
Escape character is '^]'.
dump
SessionTracker dump:
Session Sets (3):
0 expire at Wed Nov 28 20:34:00 PST 2012:
0 expire at Wed Nov 28 20:34:02 PST 2012:
1 expire at Wed Nov 28 20:34:04 PST 2012:
        0x13b4a4d22070006
ephemeral nodes dump:
Sessions with Ephemerals (0):
Connection closed by foreign host.
```

We see from this output that there is one active session. This is the session that belongs to `Master`. We also see when this session is going to expire. This expiration time is based on the session timeout that we specified when we created the `ZooKeeper` object.

Let's kill the `Master` and then repeatedly use `dump` to see the active sessions. You'll notice that it takes a while for the session to go away. This is because the server will not kill the session until the session timeout has passed. Of course, the client will continually extend the expiration time as long as it keeps an active connection to a ZooKeeper server.

When the `Master` finishes, it would be nice if its session went away immediately. This is what the `ZooKeeper.close()` method does. Once `close` is called, the session represented by the `ZooKeeper` object gets destroyed.

Let's add the `close` to our example program:

```
void stopZK() throws Exception { zk.close(); }

public static void main(String args[]) throws Exception {
    Master m = new Master(args[0]);

    m.startZK();

    // wait for a bit
    Thread.sleep(60000);

    m.stopZK();
}
```

Now we can run `Master` again and use the `dump` four-letter word to see whether the session is still active. Because `Master` explicitly closed the session, ZooKeeper did not need to wait for the session to time out before shutting it down.

## Getting Mastership

Now that we have a session, our `Master` needs to take mastership. We need to be careful, though, because there can be only one master. We also need to have multiple processes running that could become master just in case the acting master fails.

To ensure that only one master process is active at a time, we use ZooKeeper to implement the simple leader election algorithm described in "The Master Role" on page 36. In this algorithm, all potential masters try to create the `/master` znode, but only one will succeed. The process that succeeds becomes the master.

We need two things to create `/master`. First, we need the initial data for the znode. Usually we put some information about the process that becomes the master in this initial data. For now we will have each process pick a random server ID and use that for the initial data. We also need an access control list (ACL) for the new znode. Often ZooKeeper is used within a trusted environment, so an open ACL is used.

There is a constant, `ZooDefs.Ids.OPEN_ACL_UNSAFE`, that gives all permissions to everyone. (As the name indicates, this is a *very unsafe* ACL to use in untrusted environments.)

ZooKeeper provides per-znode ACLs with a pluggable authentication method, so if we need to we can restrict who can do what to which znode, but for this simple example, let's stick with `OPEN_ACL_UNSAFE`. Of course, we want the `/master` znode to go away if the acting master dies. As we saw in "Persistent and ephemeral znodes" on page 19, ZooKeeper has ephemeral znodes for this purpose. We'll define an `EPHEMERAL` znode that will automatically be deleted by ZooKeeper when the session that created it closes or is made invalid.

So, we will add the following lines to our code:

```
String serverId = Integer.toHexString(random.nextInt());

void runForMaster() {
    zk.create("/master", ❶
    serverId.getBytes(), ❷
    OPEN_ACL_UNSAFE, ❸
    CreateMode.EPHEMERAL); ❹
}
```

❶ The znode we are trying to create is `/master`. If a znode already exists with that name, the `create` will fail. We are going to store the unique ID that corresponds to this server as the data of the `/master` znode.

❷ Only byte arrays may be stored as data, so we convert the `int` to a byte array.

❸ As we mentioned, we are using an open ACL.

❹ And we are creating an `EPHEMERAL` node.

However, we aren't done yet. `create` throws two exceptions: `KeeperException` and `InterruptedException`. We need to make sure we handle these exceptions, specifically the `ConnectionLossException` (which is a subclass of `KeeperException`) and `InterruptedException`. For the rest of the exceptions we can abort the operation and move on, but in the case of these two exceptions, the `create` might have actually succeeded, so if we are the master we need to handle them.

The `ConnectionLossException` occurs when a client becomes disconnected from a ZooKeeper server. This is usually due to a network error, such as a network partition, or the failure of a ZooKeeper server. When this exception occurs, it is unknown to the client whether the request was lost before the ZooKeeper servers processed it, or if they processed it but the client did not receive the response. As we described earlier, the ZooKeeper client library will reestablish the connection for future requests, but the process must figure out whether a pending request has been processed or whether it should reissue the request.

The `InterruptedException` is caused by a client thread calling `Thread.interrupt`. This is often part of application shutdown, but it may also be used in other application-dependent ways. This exception literally interrupts the local client request processing in the process and leaves the request in an unknown state.

Because both exceptions cause a break in the normal request processing, the developer cannot assume anything about the state of the request in process. When handling these exceptions, the developer must figure out the state of the system before proceeding. In case there was a leader election, we want to make sure that we haven't been made the master without knowing it. If the `create` actually succeeded, no one else will be able to become master until the acting master dies, and if the acting master doesn't know it has mastership, no process will be acting as the master.

When handling the `ConnectionLossException`, we must find out which process, if any, has created the `/master` znode, and start acting as the leader if that process is ours. We do this by using the `getData` method:

```
byte[] getData(
    String path,
    bool watch,
    Stat stat)
```

where:

`path`
: Like with most of the other `ZooKeeper` methods, the first parameter is the path of the znode from which we will be getting the data.

`watch`
: Indicates whether we want to hear about future changes to the data returned. If set to `true`, we will get events on the `Watcher` object we set when we created the ZooKeeper handle. There is another version of this method that takes a `Watcher` object that will receive an event if changes happen. We will see how to watch for changes in future chapters, but for now we will set this parameter to `false` because we just want to know what the current data is.

`stat`
: The last parameter is a `Stat` structure that the `getData` method can fill in with metadata about the znode.

*Return value*
: If this method returns successfully (doesn't throw an exception), the byte array will contain the data of the znode.

So, let's change that code segment to the following, introducing some exception handling into our `runForMaster` method:

```
String serverId = Integer.toString(Random.nextLong());
boolean isLeader = false;

// returns true if there is a master
boolean checkMaster() {
    while (true) {
        try {
            Stat stat = new Stat();
            byte data[] = zk.getData("/master", false, stat); ❶
            isLeader = new String(data).equals(serverId)); ❷
            return true;
        } catch (NoNodeException e) {
            // no master, so try create again
            return false;
        } catch (ConnectionLossException e) {
        }
    }
}

void runForMaster() throws InterruptedException { ❸
    while (true) {
        try { ❹
            zk.create("/master", serverId.getBytes(),
                      OPEN_ACL_UNSAFE, CreateMode.EPHEMERAL); ❺
            isLeader = true;
            break;
        } catch (NodeExistsException e) {
            isLeader = false;
            break;
        } catch (ConnectionLossException e) { ❻
        }
        if (checkMaster()) break; ❼
    }
}
```

❹ We surround the `zk.create` with a `try` block so that we can handle the `ConnectionLossException`.

❺ This is the `create` request that will establish the process as the master if it succeeds.

❻ Although the body of the catch block for the ConnectionLossException is empty because we do not break, this catch will cause the process to continue to the next line.

❼ Check for an active master and retry if there is still no elected master.

❶ Check for an active master by trying to get the data for the /master znode.

❷ This line of the example shows why we need the data used in creating the /master znode: if /master exists, we use the data contained in /master to determine who the leader is. If the process received a ConnectionLossException, the current process may actually be the master; it is possible that its create request actually was processed, but the response was lost.

❸ We let the InterruptedException bubble up the call stack by letting it simply pass through to the caller.

In our example, we simply pass the InterruptedException to the caller and thus let it bubble up. Unfortunately, in Java there aren't clear guidelines for how to deal with thread interruption, or even what it means. Sometimes the interruptions are used to signal threads that things are being shut down and they need to clean up. In other cases, an interruption is used to get control of a thread, but execution of the application continues.

Our handling of InterruptedException depends on our context. If the InterruptedException will bubble up and eventually close our zk handle, we can let it go up the stack and everything will get cleaned up when the handle is closed. If the zk handle is not closed, we need to figure out if we are the master before rethrowing the exception or asynchronously continuing the operation. This latter case is particularly tricky and requires careful design to handle properly.

Our main method for the Master now becomes:

```
public static void main(String args[]) throws Exception {
    Master m = new Master(args[0]);

    m.startZK();

    m.runForMaster(); ❶

    if (isLeader) {
        System.out.println("I'm the leader"); ❷
        // wait for a bit
        Thread.sleep(60000);
    } else {
        System.out.println("Someone else is the leader");
    }

    m.stopZK();
}
```

❶ Our call to runForMaster, a method we implemented earlier, will return either when the current process has become the master or when another process is the master.

❷ Once we develop the application logic for the master, we will start executing that logic here, but for now we content ourselves with announcing victory and then waiting 60 seconds before exiting main.

Because we aren't handling the InterruptedException directly, we will simply exit the program (and therefore close our ZooKeeper handle) if it happens. Of course, the master doesn't do much before shutting down. In the next chapter, the master will actually start managing the tasks that get queued to the system, but for now let's start filling in the other components.

## Getting Mastership Asynchronously

In ZooKeeper, all synchronous calls have corresponding asynchronous calls. This allows us to issue many calls at a time from a single thread and may also simplify our implementation. Let's revisit the mastership example, this time using asynchronous calls.

Here is the asynchronous version of create:

```
void create(String path,
    byte[] data,
    List<ACL> acl,
    CreateMode createMode,
    AsyncCallback.StringCallback cb, ❶
    Object ctx) ❷
```

This version of create looks exactly like the synchronous version except for two additional parameters:

❶ An object containing the function that serves as the callback

❷ A user-specified context (an object that will be passed through to the callback when it is invoked)

This call will return immediately, usually before the create request has been sent to the server. The callback object often takes data, which we can pass through the context argument. When the result of the create request is received from the server, the context argument will be given to the application through the callback object.

Notice that this create doesn't throw any exceptions, which can simplify things for us. Because the call is not waiting for the create to complete before returning, we don't have to worry about the InterruptedException; because any request errors are encoded in the first parameter in the callback, we don't have to worry about KeeperException.

The callback object implements StringCallback with one method:

```
void processResult(int rc, String path, Object ctx, String name)
```

The asynchronous method simply queues the request to the ZooKeeper server. Transmission happens on another thread. When responses are received, they are processed on a dedicated callback thread. To preserve order, there is a single callback thread and responses are processed in the order they are received.

The parameters of processResult have the following meanings:

rc
: Returns the result of the call, which will be OK or a code corresponding to a KeeperException

path
: The path that we passed to the create

ctx
: Whatever context we passed to the create

name
: The name of the znode that was created

For now, path and name will be equal if we succeed, but if CreateMode.SEQUENTIAL is used, this will not be true.

**Callback Processing**

Because a single thread processes all callbacks, if a callback blocks, it blocks all callbacks that follow it. This means that generally you should not do intensive operations or blocking operations in a callback. There may be times when it's legitimate to use the synchronous API in a callback, but it should generally be avoided so that subsequent callbacks can be processed quickly.

So, let's start writing our master functionality. Here, we create a masterCreateCallback object that will receive the results of the create call:

```
static boolean isLeader;

static StringCallback masterCreateCallback = new StringCallback() {
    void processResult(int rc, String path, Object ctx, String name) {
        switch(Code.get(rc)) { ❶
        case CONNECTIONLOSS: ❷
            checkMaster();
            return;
        case OK: ❸
            isLeader = true;
```

```
            break;
        default: ❹
            isLeader = false;
        }
        System.out.println("I'm " + (isLeader ? "" : "not ") +
                           "the leader");
    }
};

void runForMaster() {
    zk.create("/master", serverId.getBytes(), OPEN_ACL_UNSAFE,
              CreateMode.EPHEMERAL, masterCreateCallback, null); ❺
}
```

❶ We get the result of the `create` call in `rc` and convert it to a `Code` enum to switch on. `rc` corresponds to a `KeeperException` if `rc` is not zero.

❷ If the `create` fails due to a connection loss, we will get the `CONNECTIONLOSS` result code rather than the `ConnectionLossException`. When we get a connection loss, we need to check on the state of the system and figure out what we need to do to recover. We do this in the `checkMaster` method, which we will implement next.

❸ Woohoo! We are the leader. For now we will just set `isLeader` to `true`.

❹ If any other problem happened, we did not become the leader.

❺ We kick things off in `runForMaster` when we pass the `masterCreateCallback` object to the `create` method. We pass `null` as the context object because there isn't any information that we need to pass from `runForMaster` to the `masterCreateCallback.processResult` method.

Now we have to implement the `checkMaster` method. This method looks a bit different because, unlike in the synchronous case, we string the processing logic together in the callbacks, so we do not see a sequence of events in `checkMaster`. Instead, we see things get kicked off with the `getData` method. Subsequent processing will continue in the `DataCallback` when the `getData` operation completes:

```
DataCallback masterCheckCallback = new DataCallback() {
    void processResult(int rc, String path, Object ctx, byte[] data,
                       Stat stat) {
        switch(Code.get(rc)) {
        case CONNECTIONLOSS:
            checkMaster();
            return;
        case NONODE:
            runForMaster();
            return;
        }
    }
```

```
    }

    void checkMaster() {
        zk.getData("/master", false, masterCheckCallback, null);
    }
```

The basic logic of the synchronous and asynchronous versions is the same, but in the asynchronous version we do not have a `while` loop. Instead, the error handling is done with callbacks and new asynchronous operations.

At this point the synchronous version may appear simpler to implement than the asynchronous version, but as we will see in the next chapter, our applications are often driven by asynchronous change notifications, so building everything asynchronously may result in simpler code in the end. Note that asynchronous calls also do not block the application, allowing it to make progress with other things—perhaps even submitting new ZooKeeper operations.

## Setting Up Metadata

We will use the asynchronous API to set up the metadata directories. Our master-worker design depends on three other directories: `/tasks`, `/assign`, and `/workers`. We can count on some kind of system setup to make sure everything is created before the system is started, or a master can make sure these directories are created every time it starts up. The following code segment will create these paths. There isn't really any error handling in this example apart from handling a connection loss:

```
public void bootstrap() {
    createParent("/workers", new byte[0]); ❶
    createParent("/assign", new byte[0]);
    createParent("/tasks", new byte[0]);
    createParent("/status", new byte[0]);
}

void createParent(String path, byte[] data) {
    zk.create(path,
              data,
              Ids.OPEN_ACL_UNSAFE,
              CreateMode.PERSISTENT,
              createParentCallback,
              data); ❷
}

StringCallback createParentCallback = new StringCallback() {
    public void processResult(int rc, String path, Object ctx, String name) {
        switch (Code.get(rc)) {
        case CONNECTIONLOSS:
            createParent(path, (byte[]) ctx); ❸

            break;
        case OK:
```

```
                LOG.info("Parent created");

                break;
            case NODEEXISTS:
                LOG.warn("Parent already registered: " + path);

                break;
            default:
                LOG.error("Something went wrong: ",
                          KeeperException.create(Code.get(rc), path));
            }
        }
    };
```

❶ We don't have any data to put in these znodes, so we are just passing an empty byte array.

❷ Because of that, we don't have to worry about keeping track of the data that corresponds to each znode, but often there is data unique to a path, so we will track the data using the callback context of each `create` call. It may seem a bit strange that we pass `data` in both the second and fourth parameters of `create`, but the data passed in the second parameter is the data to be written to the new znode and the data passed in the fourth will be made available to the `createParentCallback`.

❸ If the callback gets a `CONNECTIONLOSS` return code, we want to simply retry the `create`, which we can do by calling `createPath`. However, to call `createPath` we need the data that was used in the original `create`. We have that data in the `ctx` object that was passed to the callback because we passed the creation data as the fourth parameter of the `create`. Because the context object is separate from the callback object, we can use a single callback object for all of the `create`s.

In this example you will notice that there isn't any difference between a file (a znode that contains data) and a directory (a znode that contains children). Every znode can have both.

## Registering Workers

Now that we have a master, we need to set up the workers so that the master has someone to boss around. According to our design, each worker is going to create an ephemeral znode under `/workers`. We can do this quite simply with the following code. We will use the data in the znode to indicate the state of the worker:

```
import java.util.*;

import org.apache.zookeeper.AsyncCallback.DataCallback;
import org.apache.zookeeper.AsyncCallback.StringCallback;
import org.apache.zookeeper.AsyncCallback.VoidCallback;
```

```
import org.apache.zookeeper.*;
import org.apache.zookeeper.ZooDefs.Ids;
import org.apache.zookeeper.AsyncCallback.ChildrenCallback;
import org.apache.zookeeper.KeeperException.Code;
import org.apache.zookeeper.data.Stat;

import org.slf4j.*;

public class Worker implements Watcher {
    private static final Logger LOG = LoggerFactory.getLogger(Worker.class);

    ZooKeeper zk;
    String hostPort;
    String serverId = Integer.toHexString(random.nextInt());

    Worker(String hostPort) {
        this.hostPort = hostPort;
    }

    void startZK() throws IOException {
        zk = new ZooKeeper(hostPort, 15000, this);
    }

    public void process(WatchedEvent e) {
        LOG.info(e.toString() + ", " + hostPort);
    }

    void register() {
        zk.create("/workers/worker-" + serverId,
                  "Idle".getBytes(), ❶
                  Ids.OPEN_ACL_UNSAFE,
                  CreateMode.EPHEMERAL, ❷
                  createWorkerCallback, null);
    }

    StringCallback createWorkerCallback = new StringCallback() {
        public void processResult(int rc, String path, Object ctx,
                                  String name) {
            switch (Code.get(rc)) {
            case CONNECTIONLOSS:
                register(); ❸
                break;
            case OK:
                LOG.info("Registered successfully: " + serverId);
                break;
            case NODEEXISTS:
                LOG.warn("Already registered: " + serverId);
                break;
            default:
                LOG.error("Something went wrong: "
                + KeeperException.create(Code.get(rc), path));
            }
```

```
        }
    };

    public static void main(String args[]) throws Exception {
        Worker w = new Worker(args[0]);
        w.startZK();

        w.register();

        Thread.sleep(30000);
    }
}
```

❶ We will be putting the status of the worker in the data of the znode that represents the worker.

❷ If the process dies we want the znode representing the worker to get cleaned up, so we use the `EPHEMERAL` flag. That means that we can simply look at the children of `/workers` to get the list of available workers.

❸ Because this process is the only one that creates the ephemeral znode representing the process, if there is a connection loss during the creation of the znode, it can simply retry the creation.

As we have seen earlier, because we are registering an ephemeral node, if the worker dies the registered znode representing that node will go away. So this is all we need to do on the worker's side for group membership.

We are also putting status information in the znode representing the worker. This allows us to check the status of the worker by simply querying ZooKeeper. Currently, we have only the initializing and idle statuses; however, once the worker starts actually doing things, we will want to set other status information.

Here is our implementation of `setStatus`. This method works a little bit differently from methods we have seen before. We want to be able to set the status asynchronously so that it doesn't delay regular processing:

```
StatCallback statusUpdateCallback = new StatCallback() {
    public void processResult(int rc, String path, Object ctx, Stat stat) {
        switch(Code.get(rc)) {
        case CONNECTIONLOSS:
            updateStatus((String)ctx); ❶
            return;
        }
    }
};

synchronized private void updateStatus(String status) {
    if (status == this.status) { ❷
        zk.setData("/workers/" + name, status.getBytes(), -1,
                   statusUpdateCallback, status); ❸
```

```
        }
    }

    public void setStatus(String status) {
        this.status = status; ❹
        updateStatus(status); ❺
    }
```

❹ We save our status locally in case a status update fails and we need to retry.

❺ Rather than doing the update in `setStatus`, we create an `updateStatus` method that we can use in `setStatus` and in the retry logic.

❷ There is a subtle problem with asynchronous requests that we retry on connection loss: things may get out of order. ZooKeeper is very good about maintaining order for both requests and responses, but a connection loss makes a gap in that ordering, because we are creating a new request. So, before we requeue a status update, we need to make sure that we are requeuing the current status; otherwise, we just drop it. We do this check and retry in a synchronized block.

❸ We do an unconditional update (the third parameter; the expected version is –1, so version checking is disabled), and we pass the status we are setting as the context object.

❶ If we get a connection loss event, we simply need to call `updateStatus` with the status we are updating. (We passed the status in the context parameter of `setData`.) The `updateStatus` method will do checks for race conditions, so we do not need to do those here.

To understand the problems with reissuing operations on a connection loss a bit more, consider the following scenario:

1. The worker starts working on `task-1`, so it sets the status to `working on task-1`.
2. The client library tries to issue the `setData`, but encounters networking problems.
3. After the client library determines that the connection has been lost with ZooKeeper and before `statusUpdateCallback` is called, the worker finishes `task-1` and becomes idle.
4. The worker asks the client library to issue a `setData` with `Idle` as the data.
5. Then the client processes the connection lost event; if `updateStatus` does not check the current status, it would then issue a `setData` with `working on task-1`.
6. When the connection to ZooKeeper is reestablished, the client library faithfully issues the two `setData` operations in order, which means that the final state would be `working on task-1`.

By checking the current status before reissuing the `setData` in the `updateStatus` method, we avoid this scenario.

**Order and ConnectionLossException**

ZooKeeper is very strict about maintaining order and has very strong ordering guarantees. However, care should be taken when thinking about ordering in the presence of multiple threads. One common scenario where multiple threads can cause errors involves retry logic in callbacks. When reissuing a request due to a `ConnectionLossException`, a new request is created and may therefore be ordered after requests issued on other threads that occurred after the original request.

# Queuing Tasks

The final component of the system is the `Client` application that queues new tasks to be executed on a worker. We will be adding znodes under `/tasks` that represent commands to be carried out on the worker. We will be using sequential znodes, which gives us two benefits. First, the sequence number will indicate the order in which the tasks were queued. Second, the sequence number will create unique paths for tasks with minimal work. Our `Client` code looks like this:

```
import org.apache.zookeeper.ZooKeeper;
import org.apache.zookeeper.Watcher;

public class Client implements Watcher {
    ZooKeeper zk;
    String hostPort;

    Client(String hostPort) { this.hostPort = hostPort; }

    void startZK() throws Exception {
        zk = new ZooKeeper(hostPort, 15000, this);
    }

    String queueCommand(String command) throws KeeperException {
        while (true) {
            try {
                String name = zk.create("/tasks/task-", ❶
                                        command.getBytes(), OPEN_ACL_UNSAFE,
                                        CreateMode.SEQUENTIAL); ❷
                return name; ❸
                break;
            } catch (NodeExistsException e) {
                throw new Exception(name + " already appears to be running");
            } catch (ConnectionLossException e) { ❹
            }
```

```
        }

    public void process(WatchedEvent e) { System.out.println(e); }

    public static void main(String args[]) throws Exception {
        Client c = new Client(args[0]);

        c.start();

        String name = c.queueCommand(args[1]);
        System.out.println("Created " + name);
    }
}
```

❶ We are creating the znode representing a task under `/tasks`. The name will be prefixed with `task-`.

❷ Because we are using `CreateMode.SEQUENTIAL`, a monotonically increasing suffix will be appended to `task-`. This guarantees that a unique name will be created for each new task and the task ordering will be established by ZooKeeper.

❸ Because we can't be sure what the sequence number will be when we call `create` with `CreateMode.SEQUENTIAL`, the `create` method returns the name of the new znode.

❹ If we lose a connection while we have a `create` pending, we will simply retry the `create`. This may create multiple znodes for the task, causing it to be created twice. For many applications this execute-at-least-once policy may work fine. Applications that require an execute-at-most-once policy must do more work: we would need to create each of our task znodes with a unique ID (the session ID, for example) encoded in the znode name. We would then retry the `create` only if a connection loss exception happened and there were no znodes under `/tasks` with the session ID in their name.

When we run the `Client` application and pass a command, a new znode will be created in `/tasks`. It is not an ephemeral znode, so even after the `Client` application ends, any znodes it has created will remain.

## The Admin Client

Finally, we will write a simple `AdminClient` that will show the state of the system. One of the nice things about ZooKeeper is that you can use the *zkCli* utility to look at the state of the system, but usually you will want to write your own admin client to more quickly and easily administer the system. In this example, we will use `getData` and `getChildren` to get the state of our master-worker system.

These methods are simple to use. Because they don't change the state of the system, we can simply propagate errors that we encounter without having to deal with cleanup.

This example uses the synchronous versions of the calls. The methods also have a `watch` parameter that we are setting to `false` because we really want to see the current state of the system and will not be watching for changes. We will see in the next chapter how we use this parameter to track changes in the system. For now, here's our `AdminClient` code:

```
import org.apache.zookeeper.ZooKeeper;
import org.apache.zookeeper.Watcher;

public class AdminClient implements Watcher {
    ZooKeeper zk;
    String hostPort;

    AdminClient(String hostPort) { this.hostPort = hostPort; }

    void start() throws Exception {
        zk = new ZooKeeper(hostPort, 15000, this);
    }

    void listState() throws KeeperException {
        try {
            Stat stat = new Stat();
            byte masterData[] = zk.getData("/master", false, stat); ❶
            Date startDate = new Date(stat.getCtime()); ❷
            System.out.println("Master: " + new String(masterData) +
                               " since " + startDate);
        } catch (NoNodeException e) {
            System.out.println("No Master");
        }

        System.out.println("Workers:");
        for (String w: zk.getChildren("/workers", false)) {
            byte data[] = zk.getData("/workers/" + w, false, null); ❸
            String state = new String(data);
            System.out.println("\t" + w + ": " + state);
        }

        System.out.println("Tasks:");
        for (String t: zk.getChildren("/assign", false)) {
            System.out.println("\t" + t);
        }

    }

    public void process(WatchedEvent e) { System.out.println(e); }

    public static void main(String args[]) throws Exception {
        AdminClient c = new AdminClient(args[0]);
```

```
        c.start();

        c.listState();
    }
}
```

❶ We put the name of the master as the data for the `/master` znode, so getting the data for the `/master` znode will give us the name of the current master. We aren't interested in changes, which is why we pass `false` to the second parameter.

❷ We can use the information in the `Stat` structure to know how long the current master has been master. The `ctime` is the number of seconds since the epoch after which the znode was created. See `java.lang.System.currentTimeMillis()` for details.

❸ The ephemeral znode has two pieces of information: it indicates that the worker is running, and its data has the state of the worker.

The `AdminClient` is very simple: it simply runs through the data structures for our master-worker example. Try it out by starting and stopping a `Master` and some `Workers`, and running the `Client` a few times to queue up some tasks. The `AdminClient` will show the state of the system as things change.

You might be wondering if there is any advantage to using the asynchronous API in the `AdminClient`. ZooKeeper has a pipelined implementation designed to handle thousands of simultaneous requests. This is important because there are various sources of latency in the system, the largest of which are the disk and the network. Both of these components have queues that are used to efficiently use their bandwidth. The `getData` method does not do any disk access, but it must go over the network. With synchronous methods we let the pipeline drain between each request. If we are using the `AdminClient` in a relatively small system with tens of workers and hundreds of tasks, this might not be a big deal, but if we increase those numbers by an order of magnitude, the delay might become significant.

Consider a scenario in which the round trip time for a request is 1 ms. If the process needs to read 10,000 znodes, 10 seconds will be spent just on network delays. We can shorten that time to something much closer to 1 second using the asynchronous API.

Using the basic implementation of the `Master`, `Worker`, and `Client`, we have the beginnings of our master-worker system, but right now nothing is really happening. When a task gets queued, the master needs to wake up and assign the task to a worker. Workers need to find out about tasks assigned to them. Clients need to be able to know when tasks are finished. If a master fails, another master-in-waiting needs to take over. If a worker fails, its tasks need to get assigned to other workers. In upcoming chapters we will cover the concepts necessary to implement this needed functionality.

## Takeaway Messages

The commands we use in *zkCli* correspond closely to the API we use when programming with ZooKeeper, so it can be useful to do some initial experimentation with *zkCli* to try out different ways to structure application data. The API is so close to the commands in *zkCli* that you could easily write an application that matches commands used when experimenting with *zkCli*. There are some caveats, though. First, when using *zkCli* we usually are in a stable environment where unexpected failures don't happen. For code we are going to deploy, we have to handle error cases that complicate the code substantially. The `ConnectionLossException` in particular requires the developer to examine the state of the system to properly recover. (Remember, ZooKeeper helps organize distributed state and provides a framework for handling failures; it doesn't make failures go away, unfortunately.) Second, it is worthwhile to become comfortable with the asynchronous API. It can give big performance benefits and can simplify error recovery.

CHAPTER 4

# Dealing with State Change

It is not uncommon to have application processes that need to learn about changes to the state of a ZooKeeper ensemble. For instance, in our example in Chapter 1, backup masters need to know that the primary master has crashed, and workers need to know when new tasks have been assigned to them. ZooKeeper clients could, of course, poll the ZooKeeper ensemble periodically to determine whether changes have occurred. Polling, however, is not efficient, especially when the expected changes are somewhat rare.

For example, let's consider backup masters; they need to know when the primary has crashed so that they can fail over. To reduce the time it takes to recover from the primary crash, we need to poll frequently—say, every 50 ms—just for an example of aggressive polling. In this case, each backup master generates 20 requests/second. If there are multiple backup masters, we multiply this frequency by the number of backups to obtain the total request traffic generated just to poll ZooKeeper for the status of the primary master. Even if such an amount of traffic is easy for a system like ZooKeeper to deal with, primary master crashes should be rare, so most of this traffic is unnecessary. Suppose we therefore reduce the amount of polling traffic to ZooKeeper by increasing the period between requests for the status of the primary, say to 1 second. The problem with increasing this period is that it increases the time it takes to recover from a primary crash.

We can avoid this tuning and polling traffic altogether by having ZooKeeper notify interested clients of concrete events. The primary mechanism ZooKeeper provides to deal with changes is *watches*. With watches, a client registers its request to receive a one-time notification of a change to a given znode. For example, we can have the primary master create an ephemeral znode representing the master lock, and the backup masters register a watch for the existence of the master lock. In the case that the primary crashes, the master lock is automatically deleted and the backup masters are notified. Once the backup masters receive their notifications, they can start a new master election by trying to create a new ephemeral znode to represent the master lock, as we showed in "Getting Mastership" on page 51.

Watches and notifications form a general mechanism that enables clients to observe changes by other clients without having to continually poll ZooKeeper. We have illustrated with the master example, but the general mechanism is applicable to a wide variety of situations.

## One-Time Triggers

Before getting deeper into watches, let's establish some terminology. We talk about an *event* to denote the execution of an update to a given znode. A *watch* is a one-time trigger associated with a znode and a type of event (e.g., data is set in the znode, or the znode is deleted). When the watch is triggered by an event, it generates a *notification*. A notification is a message to the application client that registered the watch to inform this client of the event.

When an application process registers a watch to receive a notification, the watch is triggered at most once and upon the first event that matches the condition of the watch. For example, say that the client needs to know when a given znode `/z` is deleted (e.g., a backup master). The client executes an `exists` operation on `/z` with the watch flag set and waits for the notification. The notification comes in the form of a callback to the application client.

Each watch is associated with the session in which the client sets it. If the session expires, pending watches are removed. Watches do, however, persist across connections to different servers. Say that a ZooKeeper client disconnects from a ZooKeeper server and connects to a different server in the ensemble. The client will send a list of outstanding watches. When reregistering the watch, the server will check to see if the watched znode has changed since the watch was registered. If the znode has changed, a watch event will be sent to the client; otherwise, the watch will be reregistered at the new server.

### Wait, Can I Miss Events with One-Time Triggers?

The short answer is yes: an application can miss events between receiving a notification and registering for another watch. However, this issue deserves more discussion. Missing events is typically not a problem because any changes that have occurred during the period between receiving a notification and registering a new watch can be captured by reading the state of ZooKeeper directly.

Say that a worker receives a notification indicating that a new task has been assigned to it. To receive the new task, the worker reads the list of tasks. If several more tasks have been assigned to the worker after it received the notification, reading the list of tasks via a `getChildren` call returns all tasks. The `getChildren` call also sets a new watch, guaranteeing that the worker will not miss tasks.

Actually, having one notification amortized across multiple events is a positive aspect. It makes the notification mechanism much more lightweight than sending a notification for every event for applications that have a high rate of updates. To give a trivial example, if every notification captures two events on average, we are generating only 0.5 notifications per event instead of 1 notification per event.

## Getting More Concrete: How to Set Watches

All read operations in the ZooKeeper API—`getData`, `getChildren`, and `exists`—have the option to set a watch on the znode they read. To use the watch mechanism, we need to implement the `Watcher` interface, which consists of implementing a `process` method:

```
public void process(WatchedEvent event);
```

The `WatchedEvent` data structure contains the following:

- The state of the ZooKeeper session (`KeeperState`): `Disconnected`, `SyncConnected`, `AuthFailed`, `ConnectedReadOnly`, `SaslAuthenticated`, or `Expired`
- The event type (`EventType`): `NodeCreated`, `NodeDeleted`, `NodeDataChanged`, `NodeChildrenChanged`, or `None`
- A znode path in the case that the watched event is not `None`

The first three events refer to a single znode, whereas the fourth event concerns the children of the znode on which it is issued. We use `None` when the watched event is for a change of the state of the ZooKeeper session.

There are two types of watches: data watches and child watches. Creating, deleting, or setting the data of a znode successfully triggers a data watch. Both `exists` and `getData` set data watches. Only `getChildren` sets child watches, which are triggered when a child znode is either created or deleted. For each event type, we have the following calls for setting a watch:

`NodeCreated`
: A watch is set with a call to `exists`.

`NodeDeleted`
: A watch is set with a call to either `exists` or `getData`.

`NodeDataChanged`
: A watch is set with either `exists` or `getData`.

`NodeChildrenChanged`
: A watch is set with `getChildren`.

When creating a `ZooKeeper` object (see Chapter 3), we need to pass a default `Watcher` object. The ZooKeeper client uses this watcher to notify the application of changes to the ZooKeeper state, in case the state of the session changes. For event notifications related to ZooKeeper znodes, you can either use the default watcher or implement a different one. For example, the `getData` call has two different ways of setting a watch:

```
public byte[] getData(final String path, Watcher watcher, Stat stat);
public byte[] getData(String path, boolean watch, Stat stat);
```

Both signatures pass the znode as the first argument. The first signature passes a new `Watcher` object (which we must have created). The second signature tells the client to use the default watcher, and only requires `true` as the second parameter of the call.

The `stat` input parameter is an instance of the `Stat` structure that ZooKeeper uses to return information about the znode designated by `path`. The `Stat` structure contains information about the znode, such as the timestamp of the last change (zxid) that changed this znode and the number of children in the znode.

One important observation about watches is that currently it is not possible to remove them once set. The only two ways to remove a watch are to have it triggered or for its session to be closed or expired. This behavior is likely to change in future versions, however, because the community has been working on it for version 3.5.0.

**A Bit of Overloading**

We use the same watch mechanism for notifying the application of events related to the state of a ZooKeeper session and events related to znode changes. Although session state changes and znode state changes constitute independent sets of events, we rely upon the same mechanism to deliver such events for simplicity.

## A Common Pattern

Before we get into some snippets for the master-worker example, let's take a quick look at a pretty common code pattern used in ZooKeeper applications:

1. Make an asynchronous call.
2. Implement a callback object and pass it to the asynchronous call.
3. If the operation requires setting a watch, then implement a `Watcher` object and pass it on to the asynchronous call.

A code sample for this pattern using an asynchronous `exists` call looks like this:

```
zk.exists("/myZnode", ❶
          myWatcher,
          existsCallback,
          null);
```

```
Watcher myWatcher = new Watcher() { ❷
    public void process(WatchedEvent e) {
        // Process the watch event
    }
}

StatCallback existsCallback = new StatCallback() { ❸
    public void processResult(int rc, String path, Object ctx, Stat stat) {
        // Process the result of the exists call
    }
};
```

❶ ZooKeeper `exists` call. Note that the call is asynchronous.

❷ Watcher implementation.

❸ `exists` callback.

As we will see next, we'll make extensive use of this skeleton.

# The Master-Worker Example

Let's now look at how we deal with changes of state in the master-worker example. Here is a list of tasks that require a component to wait for changes:

- Mastership changes
- Master waits for changes to the list of workers
- Master waits for new tasks to assign
- Worker waits for new task assignments
- Client waits for task execution result

We next show some code snippets to illustrate how to code these tasks with ZooKeeper. We provide the complete example code as part of the additional material to this book.

## Mastership Changes

Recall from "Getting Mastership" on page 51 that an application client elects itself master by creating the `/master` znode (we call this "running for master"). If the znode already exists, the application client determines that it is not the primary master and returns. That implementation, however, does not tolerate a crash of the primary master. If the primary master crashes, the backup masters won't know about it. Consequently, we need to set a watch on `/master` so that ZooKeeper notifies the client when `/master` is deleted (either explicitly or because the session of the primary master has expired).

To set the watch, we create a new watcher named `masterExistsWatcher` and pass it to `exists`. Upon a notification of `/master` being deleted, the `process` call defined in `masterExistsWatcher` calls `runForMaster`:

```
StringCallback masterCreateCallback = new StringCallback() {
    public void processResult(int rc, String path, Object ctx, String name) {
        switch (Code.get(rc)) {
        case CONNECTIONLOSS:
            checkMaster(); ❶

            break;
        case OK:
            state = MasterStates.ELECTED;
            takeLeadership(); ❷

            break;
        case NODEEXISTS:
            state = MasterStates.NOTELECTED;
            masterExists(); ❸

            break;
        default:
            state = MasterStates.NOTELECTED;
            LOG.error("Something went wrong when running for master.", ❹
                      KeeperException.create(Code.get(rc), path));
        }
    }
};

void masterExists() {
    zk.exists("/master",
              masterExistsWatcher, ❺
              masterExistsCallback,
              null);
}

Watcher masterExistsWatcher = new Watcher() {
    public void process(WatchedEvent e) {
        if(e.getType() == EventType.NodeDeleted) {
            assert "/master".equals( e.getPath() );

            runForMaster(); ❻
        }
    }
};
```

❶ In the case of a connection loss event, the client checks if the `/master` znode is there because it doesn't know if it has been able to create it or not.

❷ If `OK`, then it simply takes leadership.

❸ If someone else has already created the znode, then the client needs to watch it.

❹ If anything unexpected happens, then it logs the error and doesn't do anything else.

❺ This `exists` call is to set a watch on the `/master` znode.

❻ If the `/master` znode is deleted, then it runs for master again.

Following the asynchronous style we used in "Getting Mastership Asynchronously" on page 56, we also create a callback method for the `exists` call that takes care of a few cases. First, it tries the `exists` operation again in the case of a connection loss event because it needs to set a watch on `/master`. Second, it is possible for the `/master` znode to be deleted between the execution of the `create` callback and the execution of the `exists` operation. Consequently, we need to check whether `stat` is null whenever the response from `exists` is positive (return code is `OK`). `stat` is null when the node does not exist. Third, if the return is not `OK` or `CONNECTIONLOSS`, then we check for the `/master` znode by getting its data. Say that the client session expires. In this case, the callback to get the data of `/master` logs an error message and exits. Our `exists` callback looks like this:

```
StatCallback masterExistsCallback = new StatCallback() {
    public void processResult(int rc, String path, Object ctx, Stat stat) {
        switch (Code.get(rc)) {
        case CONNECTIONLOSS:
            masterExists(); ❶

            break;
        case OK:
            if(stat == null) {
                state = MasterStates.RUNNING;
                runForMaster(); ❷
            }

            break;
        default:
            checkMaster(); ❸
            break;
        }
    }
};
```

❶ In the case of a connection loss, just try again.

❷ If it returns `OK`, run for master only in the case that the znode is gone.

❸ If something unexpected happens, check if `/master` is there by getting its data.

The result of the exists operation over `/master` might be that the znode has been deleted. In this case, the client needs to run for `/master` again because it is not guaranteed that the watch was set before the znode was deleted. If the new attempt to become primary fails, then the client knows that some other client succeeded and it tries to watch `/master` again. In the case the notification for `/master` indicates that it has been created instead of deleted, the client does not run for `/master`. At the same time, the corresponding exists operation (the one that has set the watch) must have returned that `/master` doesn't exist, which triggers the procedure to run for `/master` from the exists callback.

Note that this pattern of running for master and executing `exists` to set a watch on `/master` continues for as long as the client runs and does not become a primary master. If it becomes the primary master and eventually crashes, the client can restart and reexecute this code.

Figure 4-1 makes the possible interleavings of operations more explicit. If the `create` operation executed when running for primary master succeeds (a), the application client doesn't have to do anything else. If the `create` operation fails, meaning that the node already exists, the client executes an `exists` operation to set a watch on the `/master` znode (b). Between running for master and executing the `exists` operation, it is possible that the `/master` znode gets deleted. In this case, the `exists` call indicates that the znode still exists, and the client simply waits for a notification. Otherwise, it tries to run for master again, trying to create the `/master` znode. If creating the `/master` znode succeeds, the watch is triggered, indicating that there has been a change to the znode (c). This notification, however, is meaningless, because the client itself made the change. If the `create` fails again, we restart the process by executing `exists` and setting a watch on `/master` (d).

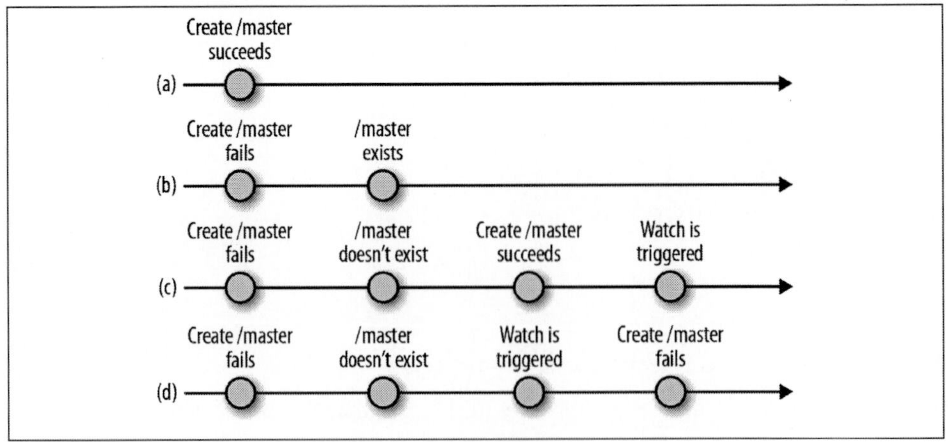


*Figure 4-1. Running for master, possible interleavings*

## Master Waits for Changes to the List of Workers

New workers may be added to the system and old workers may be decommissioned at any time. Workers might also crash before executing their assignments. To determine the workers that are available at any one time, we register new workers with ZooKeeper by adding a znode as a child of `/workers`. When a worker crashes or is simply removed from the system, its session expires, automatically causing its znode to be removed. Polite workers may also explicitly close their sessions without making ZooKeeper wait for a session expiration.

The primary master uses `getChildren` to obtain the list of available workers and to watch for changes to the list. Sample code to obtain the list and watch for changes is as follows:

```
Watcher workersChangeWatcher = new Watcher() { ❶
    public void process(WatchedEvent e) {
        if(e.getType() == EventType.NodeChildrenChanged) {
            assert "/workers".equals( e.getPath() );

            getWorkers();
        }
    }
};

void getWorkers() {
    zk.getChildren("/workers",
                    workersChangeWatcher,
                    workersGetChildrenCallback,
                    null);
}

ChildrenCallback workersGetChildrenCallback = new ChildrenCallback() {
    public void processResult(int rc, String path, Object ctx,
                              List<String> children) {
        switch (Code.get(rc)) {
        case CONNECTIONLOSS:
            getWorkerList(); ❷

            break;
        case OK:
            LOG.info("Succesfully got a list of workers: "
                     + children.size()
                     + " workers");
            reassignAndSet(children); ❸

            break;
        default:
            LOG.error("getChildren failed",
                      KeeperException.create(Code.get(rc), path));
        }
```

```
    }
};
```

❶ workersChangeWatcher is the watcher for the list of workers.

❷ In the case of a CONNECTIONLOSS event, we need to reexecute the operation to obtain the children and set the watch.

❸ This call reassigns tasks of dead workers and sets the new list of workers.

We start by calling getWorkerList. This call executes getChildren asynchronously, passing workersGetChildrenCallback to process the result of the operation. If the client disconnects from a server (CONNECTIONLOSS event), the watch is not set and we don't have a list of workers; we execute getWorkerList again to set the watch and obtain the list of workers. Upon a successful execution of getChildren, we call reassignAndSet, as in the following code:

```
ChildrenCache workersCache; ❶

void reassignAndSet(List<String> children) {
    List<String> toProcess;

        if(workersCache == null) {
            workersCache = new ChildrenCache(children); ❷
            toProcess = null; ❸
        } else {
            LOG.info( "Removing and setting" );
            toProcess = workersCache.removedAndSet( children ); ❹
        }

        if(toProcess != null) {
            for(String worker : toProcess) {
                getAbsentWorkerTasks(worker); ❺
            }
        }
    }
```

❶ Here's the cache that holds the last set of workers we have seen.

❷ If this is the first time it is using the cache, then instantiate it.

❸ The first time we get workers, there is nothing to do.

❹ If it is not the first time, then we need to check if some worker has been removed.

❺ If there is any worker that has been removed, then we need to reassign its tasks.

We use the cache because we need to remember what we have seen before. Say that we get the list of workers for the first time. When we get the notification that the list of workers has changed, we won't know what exactly has changed even after reading it again unless we keep the old values. The cache class for this example simply keeps the last list the master has read and implements a couple of methods to determine what has changed.

**Watch upon CONNECTIONLOSS**

A watch for a znode is set only if the operation is successful. If the ZooKeeper operation fails to be executed because the client has disconnected, then the application needs to call it again.

## Master Waits for New Tasks to Assign

Like waiting for changes to the list of workers, the primary master waits for new tasks to be added to `/tasks`. The master initially obtains the set of current tasks and sets a watch for changes to the set. The set is represented in ZooKeeper by the children of `/tasks`, and each child corresponds to a task. Once the master obtains tasks that have not been assigned, it selects a worker at random and assigns the task to the worker. We implement the assignment in `assignTasks`:

```
Watcher tasksChangeWatcher = new Watcher() { ❶
    public void process(WatchedEvent e) {
        if(e.getType() == EventType.NodeChildrenChanged) {
            assert "/tasks".equals( e.getPath() );

            getTasks();
        }
    }
};

void getTasks() {
    zk.getChildren("/tasks",
                   tasksChangeWatcher,
                   tasksGetChildrenCallback,
                   null); ❷
}

ChildrenCallback tasksGetChildrenCallback = new ChildrenCallback() {
    public void processResult(int rc,
                              String path,
                              Object ctx,
                              List<String> children) {
        switch(Code.get(rc)) {
        case CONNECTIONLOSS:
            getTasks();

            break;
        case OK:
            if(children != null) {
                assignTasks(children); ❸
            }
```

```
                break;
            default:
                LOG.error("getChildren failed.",
                          KeeperException.create(Code.get(rc), path));
            }
        }
    };
```

❶ Watcher implementation to handle a notification that the list of tasks has changed.

❷ Get the list of tasks.

❸ Assign tasks in the list.

Now we'll implement `assignTasks`. It simply assigns each of the tasks in the list of children of `/tasks`. Before creating the assignment znode, we get the task data with `getData`:

```
void assignTasks(List<String> tasks) {
    for(String task : tasks) {
        getTaskData(task);
    }
}

void getTaskData(String task) {
    zk.getData("/tasks/" + task,
               false,
               taskDataCallback,
               task); ❶
}

DataCallback taskDataCallback = new DataCallback() {
    public void processResult(int rc,
                              String path,
                              Object ctx,
                              byte[] data,
                              Stat stat)  {
        switch(Code.get(rc)) {
        case CONNECTIONLOSS:
            getTaskData((String) ctx);

            break;
        case OK:
            /*
             * Choose worker at random.
             */
            int worker = rand.nextInt(workerList.size());
            String designatedWorker = workerList.get(worker);

            /*
```

```
                 * Assign task to randomly chosen worker.
                 */
                String assignmentPath = "/assign/" + designatedWorker + "/" +
                                         (String) ctx;
                createAssignment(assignmentPath, data); ❷

                break;
            default:
                LOG.error("Error when trying to get task data.",
                        KeeperException.create(Code.get(rc), path));
            }
        }
    };
```

❶ Get task data.

❷ Select a worker randomly and assign the task to this worker.

We need to get the task data first because we delete the task znode under `/tasks` after assigning it. This way the master doesn't have to remember which tasks it has assigned. Let's look at the code for assigning a task:

```
void createAssignment(String path, byte[] data) {
    zk.create(path,
            data, Ids.OPEN_ACL_UNSAFE,
            CreateMode.PERSISTENT,
            assignTaskCallback,
            data); ❶
}

StringCallback assignTaskCallback = new StringCallback() {
    public void processResult(int rc, String path, Object ctx, String name) {
        switch(Code.get(rc)) {
        case CONNECTIONLOSS:
            createAssignment(path, (byte[]) ctx);

            break;
        case OK:
            LOG.info("Task assigned correctly: " + name);
            deleteTask(name.substring( name.lastIndexOf("/") + 1 )); ❷

            break;
        case NODEEXISTS:
            LOG.warn("Task already assigned");

            break;
        default:
            LOG.error("Error when trying to assign task.",
                    KeeperException.create(Code.get(rc), path));
        }
    }
};
```

❶ Create an assignment. The path is of the form /assign/worker-id/task-num.

❷ Delete the task znode under /tasks.

For new tasks, after the master selects a worker to assign the task to, it creates a znode under /assign/worker-id, where id is the identifier of the worker. Next, it deletes the znode from the list of pending tasks. The code for deleting the znode in the previous example follows the pattern of earlier code we have shown.

When the master creates an assignment znode for a worker with identifier id, ZooKeeper generates a notification for the worker, assuming that the worker has a watch registered upon its assignment znode (/assign/worker-id).

Note that the master also deletes the task znode under /tasks after assigning it successfully. This approach simplifies the role of the master when it receives new tasks to assign. If the list of tasks mixed the assigned and unassigned tasks, the master would need a way to disambiguate the tasks.

## Worker Waits for New Task Assignments

One of the first steps a worker has to execute is to register itself with ZooKeeper. It does this by creating a znode under /workers, as we already discussed:

```
void register() {
        zk.create("/workers/worker-" + serverId,
                  new byte[0],
                  Ids.OPEN_ACL_UNSAFE,
                  CreateMode.EPHEMERAL,
                  createWorkerCallback, null); ❶
}

StringCallback createWorkerCallback = new StringCallback() {
    public void processResult(int rc, String path, Object ctx, String name) {
        switch (Code.get(rc)) {
        case CONNECTIONLOSS:
            register(); ❷

            break;
        case OK:
            LOG.info("Registered successfully: " + serverId);

            break;
        case NODEEXISTS:
            LOG.warn("Already registered: " + serverId);

            break;
        default:
            LOG.error("Something went wrong: " +
                      KeeperException.create(Code.get(rc), path));
        }
```

```
        }
    };
```

1. Register the worker by creating a znode.
2. Try again. Note that registering again is not a problem. If the znode has already been created, we get a `NODEEXISTS` event back.

Adding this znode signals to the master that this worker is active and ready to process tasks. Note that we don't use here the idle/busy status introduced in Chapter 3 to simplify the example.

We similarly create a znode `/assign/worker-id` so that the master can assign tasks to this worker. If we create `/workers/worker-id` before `/assign/worker-id`, we could fall into the situation in which the master tries to assign the task but cannot because the assigned parent's znode has not been created yet. To avoid this situation, we need to create `/assign/worker-id` first. Moreover, the worker needs to set a watch on `/assign/worker-id` to receive a notification when a new task is assigned.

Once the worker has the list of tasks assigned to it, it fetches the tasks from `/assign/worker-id` and executes them. The worker takes each task in its list and verifies whether it has already queued the task for execution. It keeps a list of ongoing tasks for this purpose. Note that we loop through the assigned tasks of a worker in a separate thread to release the callback thread. Otherwise, we would be blocking other incoming callbacks. In our example, we use a Java `ThreadPoolExecutor` to allocate a thread to loop through the tasks:

```
Watcher newTaskWatcher = new Watcher() {
    public void process(WatchedEvent e) {
        if(e.getType() == EventType.NodeChildrenChanged) {
            assert new String("/assign/worker-"+ serverId).equals( e.getPath() );

            getTasks(); ❶
        }
    }
};

void getTasks() {
    zk.getChildren("/assign/worker-" + serverId,
                   newTaskWatcher,
                   tasksGetChildrenCallback,
                   null);
}

ChildrenCallback tasksGetChildrenCallback = new ChildrenCallback() {
    public void processResult(int rc,
                              String path,
                              Object ctx,
                              List<String> children) {
```

```
        switch(Code.get(rc)) {
        case CONNECTIONLOSS:
            getTasks();
            break;
        case OK:
            if(children != null) {
                executor.execute(new Runnable() { ❷
                    List<String> children;
                    DataCallback cb;

                    public Runnable init (List<String> children,
                                          DataCallback cb) {
                        this.children = children;
                        this.cb = cb;

                        return this;
                    }

                    public void run() {
                        LOG.info("Looping into tasks");
                        synchronized(onGoingTasks) {
                            for(String task : children) { ❸
                                if(!onGoingTasks.contains( task )) {
                                    LOG.trace("New task: {}", task);
                                    zk.getData("/assign/worker-" +
                                               serverId + "/" + task,
                                               false,
                                               cb,
                                               task); ❹
                                    onGoingTasks.add( task ); ❺
                                }
                            }
                        }
                    }
                }
                .init(children, taskDataCallback));
            }
            break;
        default:
            System.out.println("getChildren failed: " +
                               KeeperException.create(Code.get(rc), path));
        }
    }
};
```

❶ Upon receiving a notification that the children have changed, get the list of children.

❷ Execute in a separate thread.

❸ Loop through the list of children.

❹ Get task data to execute it.

❺ Add task to the list of tasks being executed to avoid executing it multiple times.

**Session Events and Watchers**

When we disconnect from a server (for example, when the server crashes), no watches are delivered until the connection is reestablished. For this reason, session events like `CONNECTIONLOSS` are sent to all outstanding watch handlers. In general, applications use session events to go into a safe mode: the ZooKeeper client does not receive events while disconnected, so it should act conservatively in this state. In the case of our toy master-worker application, all actions except submitting a task are reactive, so if a master or a worker is disconnected, it simply does not trigger any action. Also, the master-worker client is not able to submit new tasks and it does not receive status notifications while disconnected.

## Client Waits for Task Execution Result

Suppose an application client has submitted a task. Now it needs to know when it has been executed and its status. Recall that once a worker executes a task, it creates a znode under `/status`. Let's first check the code to submit a task for execution:

```
void submitTask(String task, TaskObject taskCtx) {
    taskCtx.setTask(task);
    zk.create("/tasks/task-",
              task.getBytes(),
              Ids.OPEN_ACL_UNSAFE,
              CreateMode.PERSISTENT_SEQUENTIAL,
              createTaskCallback,
              taskCtx); ❶
}

StringCallback createTaskCallback = new StringCallback() {
    public void processResult(int rc, String path, Object ctx, String name) {
        switch (Code.get(rc)) {
        case CONNECTIONLOSS:
            submitTask(((TaskObject) ctx).getTask(),
                       (TaskObject) ctx); ❷

            break;
        case OK:
            LOG.info("My created task name: " + name);
            ((TaskObject) ctx).setTaskName(name);
            watchStatus("/status/" + name.replace("/tasks/", ""),
                       ctx); ❸

            break;
        default:
```

```
                LOG.error("Something went wrong" +
                          KeeperException.create(Code.get(rc), path));
            }
        }
    };
```

❶ Unlike previous calls to ZooKeeper, we are passing a context object, which is an instance of a `Task` class in our implementation.

❷ Resubmit the task upon a connection loss. Note that resubmitting may create a duplicate of the task.

❸ Set a watch on the status znode for this task.

**Has My Sequential Znode Been Created?**

Dealing with a `CONNECTIONLOSS` event when trying to create a sequential znode is somewhat tricky. Because ZooKeeper assigns the sequence number, it is not possible for the disconnected client to determine whether the znode has been created when there might be concurrent requests to create sequential znodes from other clients. (Note that all the create requests we are talking about in this note refer to the children of the same znode.)

To overcome this limitation, we have to give some hint about the originator of the znode, like adding the server ID as part of the task name. Using this approach, it is possible to determine whether the task has been created by listing all task znodes.

Here we check if the status node already exists (maybe the task has been processed fast) and set a watch. We provide a watcher implementation to react to the notification of the znode creation and a callback implementation for the `exists` call:

```
ConcurrentHashMap<String, Object> ctxMap =
    new ConcurrentHashMap<String, Object>();

void watchStatus(String path, Object ctx) {
    ctxMap.put(path, ctx);
    zk.exists(path,
              statusWatcher,
              existsCallback,
              ctx); ❶
}

Watcher statusWatcher = new Watcher() {
    public void process(WatchedEvent e) {
        if(e.getType() == EventType.NodeCreated) {
            assert e.getPath().contains("/status/task-");

            zk.getData(e.getPath(),
```

```
                                    false,
                                    getDataCallback,
                                    ctxMap.get(e.getPath()));
                }
            }
        };

        StatCallback existsCallback = new StatCallback() {
            public void processResult(int rc, String path, Object ctx, Stat stat) {
                switch (Code.get(rc)) {
                case CONNECTIONLOSS:
                    watchStatus(path, ctx);

                    break;
                case OK:
                    if(stat != null) {
                        zk.getData(path, false, getDataCallback, null); ❷
                    }

                    break;
                case NONODE:
                    break; ❸
                default:
                    LOG.error("Something went wrong when " +
                              "checking if the status node exists: " +
                              KeeperException.create(Code.get(rc), path));

                    break;
                }
            }
        };
```

❶ The client propagates the context object here so that it can modify the task object (`TaskObject`) accordingly when it receives a notification for the status znode.

❷ The status znode is already there, so the client needs to get it.

❸ If the status znode is not there yet, which should typically be the case, the client does nothing.

## An Alternative Way: Multiop

Multiop was not in the original design of ZooKeeper, but was added in version 3.4.0. Multiop enables the execution of multiple ZooKeeper operations in a block atomically. The execution is *atomic* in the sense that either all operations in a multiop block succeed or all fail. For example, we can delete a parent znode and its child in a multiop block. The only possible outcomes are that either both operations succeed or both fail. It is not possible for the parent to be deleted while leaving one of its children around, or vice versa.

To use the multiop feature:

1. Create an `Op` object to represent each ZooKeeper operation you intend to execute through a multiop call. ZooKeeper provides an `Op` implementation for each of the operations that change state: `create`, `delete`, and `setData`.
2. Within the `Op` object, call a static method provided by `Op` for that operation.
3. Add this `Op` object to an `Iterable` Java object, such as a list.
4. Call `multi` on the list.

The following example illustrates this process:

```
Op deleteZnode(String z) { ❶
    return Op.delete(z, -1); ❷
}

...

List<OpResult> results = zk.multi(Arrays.asList(deleteZnode("/a/b"),
                                  deleteZnode("/a")); ❸
```

❶ Create an `Op` object for the `delete` call.

❷ Return the object by calling the appropriate `Op` method.

❸ Execute both `delete` calls as a unit by using the `multi` call and passing them as a list of `Op` instances.

The call to `multi` returns a list of `OpResult` objects, each one specialized to the corresponding operation. For example, for the `delete` operation we have a `DeleteResult` class, which extends `OpResult`. The methods and data exposed by each result object depend on the operation type. `DeleteResult` offers only `equals` and `hashCode` methods, whereas `CreateResult` exposes the path of the operation and a `Stat` object. In the presence of errors, ZooKeeper returns an instance of `ErrorResult` containing an error code.

The `multi` call also has an asynchronous version. Here are the signatures of the synchronous and asynchronous methods:

```
public List<OpResult> multi(Iterable<Op> ops) throws InterruptedException,
                                                   KeeperException;
public void multi(Iterable<Op> ops, MultiCallback cb, Object ctx);
```

`Transaction` is a wrapper for `multi` with a simpler interface. We can create an instance of `Transaction`, add operations, and commit the transaction. The previous example rewritten using `Transaction` looks like this:

```
Transaction t = new Transaction();
t.delete("/a/b", -1);
```

```
t.delete("/a", -1);
List<OpResult> results = t.commit();
```

The `commit` call also has an asynchronous version that takes as input a `MultiCallback` object and a context object:

```
public void commit(MultiCallback cb, Object ctx);
```

`multiop` can simplify our master-worker implementation in at least one place. When assigning a task, the master in previous examples has created the corresponding assignment znode and then deleted the task znode under `/tasks`. If the master crashes before deleting the znode under `/tasks`, we are left with a task in `/tasks` that has already been assigned. Using `multiop`, we can create the znode representing the assignment of the task under `/assign` and delete the znode representing the task under `/tasks` atomically. Using this approach, we guarantee that no task znode under `/tasks` has been already assigned. If a backup takes over the role of master, it is not necessary to disambiguate the tasks in `/tasks`: they are all unassigned.

Another feature that `multiop` offers is the possibility of checking the version of a znode to enable operations over multiple znodes that read the state of ZooKeeper and write back some data—possibly a modification of what has been read. The version of the znode that is checked does not change, so this call enables a multiop that checks the version of a znode that is not modified. This feature is useful when the changes to one or more znodes are conditional upon the version of another znode. Say that in our master-worker example, the master needs to have the clients adding new tasks under a path that the master specifies. For example, the master could ask clients to create new tasks as children of `/tasks-`*`mid`*, where *`mid`* is the master identifier. The master stores this path as the data of the `/master-path` znode. A client that needs to add a new task first reads `/master-path` and picks its current version with `Stat`. Next, the client creates a new task znode under `/tasks-`*`mid`* as part of the a `multiop` call, and it also checks that the version of `/master-path` matches the one it has read.

The signature of `check` is similar to that of `setData`, but it doesn't include data:

```
public static Op check(String path, int version);
```

If the version of the znode in the given `path` does not match, the `multi` call fails. To illustrate, this is roughly how the code would look if we were to implement the example we have just discussed:

```
byte[] masterData = zk.getData("/master-path", false, stat); ❶
String parent = new String(masterData); ❷
...

zk.multi(Arrays.asList(Op.check("/master-path", stat.getVersion()),
                       Op.create(, modify(z1Data),-1), ❸
```

❶ Get the data of `/master`.

❷ Extract the path from the `/master` znode.

❸ `multi` call with two operations.

Note that if we store the path along with the master ID in `/master`, this scheme does not work. The `/master` znode is created every time by a new master, which makes its version consistently `1`.

## Watches as a Replacement for Explicit Cache Management

It is undesirable from the application's perspective to have clients accessing ZooKeeper every time they need to get the data for a given znode, the list of children of a znode, or anything else related to the ZooKeeper state. Instead, it is much more efficient to have clients cache values locally and use them at will. Once such values change, of course, you want ZooKeeper to notify the clients so they can update the caches. These notifications are the same ones that we have been talking about to this point, and as before, application clients register to receive such notifications through watches. In short, these watches enable clients to cache a local version of a value (such as the data of a znode or its list of children) and to receive notifications when that value changes.

An alternative to the approach that the ZooKeeper designers have adopted would be to cache transparently on behalf of the client all ZooKeeper state it accesses and to invalidate the values transparently when there are updates to cached data. Implementing such a cache coherence scheme could be costly, however, because clients might not need to cache all ZooKeeper state they access, and servers would need to invalidate cached state nonetheless. To implement invalidation, servers would have to either keep track of the cache content for each client or broadcast invalidation requests. Both options are expensive for a large number of clients and undesirable from our perspective.

Regardless of which party manages the client cache—ZooKeeper directly or the ZooKeeper application—notifying clients of updates can be performed either synchronously or asynchronously. Synchronously invalidating state across all clients holding a copy would be inefficient, because clients often proceed at different paces and consequently slow clients would force other clients to wait. Such differences become more frequent as the size of the client population increases.

The notifications approach that the designers opted for can be perceived as an asynchronous way of invalidating ZooKeeper state on the client side. ZooKeeper queues notifications to clients, and such notifications are consumed asynchronously. This invalidation scheme is also optional; it is up to the application to decide what parts of the ZooKeeper state require invalidation for any given client. These design choices are a better match for the use cases of ZooKeeper.

# Ordering Guarantees

There are a few important observations to keep in mind with respect to ordering when implementing applications with ZooKeeper.

## Order of Writes

ZooKeeper state is replicated across all servers forming the ensemble of an installation. The servers agree upon the order of state changes and apply them using the same order. For example, if a ZooKeeper server applies a state change that creates a znode `/z` followed by a state change that deletes a znode `/z'`, all servers in the ensemble must also apply these changes, and in the same order.

Servers, however, do not necessarily apply state updates simultaneously. In fact, they rarely do. Servers most likely apply state changes at different times because they proceed at different speeds, even if the hardware they run upon is fairly homogeneous. There are a number of reasons that could cause this time lag, such as operating system scheduling and background tasks.

Applying state updates at different times is typically not a problem for applications because they still perceive the same order of updates. Applications may perceive it, however, if ZooKeeper state is communicated through hidden channels, as we discuss next.

## Order of Reads

ZooKeeper clients always observe the same order of updates, even if they are connected to different servers. But it is possible for two clients to observe updates at different times. If they communicate outside ZooKeeper, the difference becomes apparent.

Let's consider the following situation:

- A client $c_1$ updates the data of a znode `/z` and receives an acknowledgment.
- Client $c_1$ sends a message through a direct TCP connection to a client $c_2$ saying that it has changed the state of `/z`.
- Client $c_2$ reads the state of `/z` but observes a state previous to the update of $c_1$.

We call this a *hidden channel* because ZooKeeper doesn't know about the clients' extra communication. Now $c_2$ has stale data. This situation is illustrated in Figure 4-2.

To avoid reading stale data, we advise that applications use ZooKeeper for all communication related to the ZooKeeper state. For example, to avoid the situation just described, $c_2$ could set a watch on `/z` instead of receiving a direct message from $c_1$. With a watch, $c_2$ learns of the change to `/z` and eliminates the hidden channel problem.

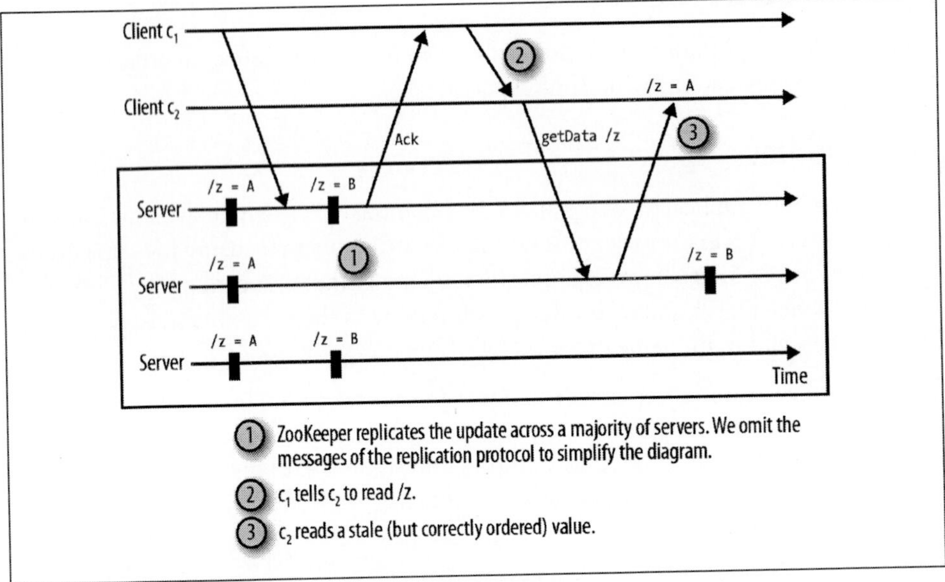


*Figure 4-2. Example of the hidden channel problem*

## Order of Notifications

ZooKeeper orders notifications with respect to other notifications and asynchronous replies, respecting the order of updates to the system state. Say that ZooKeeper orders two state updates *u* and *u′*, with *u′* following *u*. If *u* and *u′* modify znodes `/a` and `/b`, respectively, and a client *c* has a watch set on `/a`, *c* is able to observe the update *u′* only by, say, reading `/b` after receiving the notification corresponding to *u*.

This ordering enables applications to use watches to implement safety properties. Say that a znode `/z` is created or deleted to indicate that some configuration stored in ZooKeeper is invalid. Guaranteeing that clients are notified of the creation or deletion of `/z` before any actual change is made to the configuration is important to make sure that clients won't read an invalid configuration.

To make it more concrete, say that we have a znode `/config` that is the parent of a number of other znodes containing application configuration metadata: `/config/m1`, `/config/m2`, ..., `/config/m_n_`. For the purposes of this example, it doesn't matter what the content of the znodes actually is. Say that a master application process needs to update these nodes by invoking `setData` on each znode, and it can't have a client reading a partial update to these znodes. One solution is to have the master create a `/config/invalid` znode before it starts updating the configuration znodes. Other clients that need to read this state watch `/config/invalid` and avoid reading it if the invalid znode is present. Once the invalid znode is deleted, meaning that a new valid set of configuration znodes is available, clients can proceed to read that set.

For this particular example, we could alternatively have used `multiop` to execute all `setData` operations to the `/config/m[1-n]` znodes atomically instead of using a znode to mark some state as partially modified. In instances in which atomicity is the problem, we can use `multiop` instead of relying upon an extra znode and notifications. The notification mechanism, however, is more general and is not constrained to atomicity.

Because ZooKeeper orders notifications according to the order of the state updates that trigger the notifications, clients can rely upon perceiving the true order of ZooKeeper state changes through their notifications.

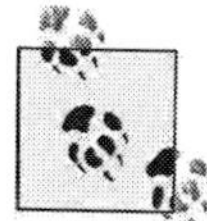

**Liveness versus Safety**

We have used the notifications mechanism extensively for liveness in this chapter. *Liveness* is about making sure that the system eventually makes progress. Notifications of new tasks and new workers are examples of events related to liveness. If a master is not notified of a new task, the task will never be executed. Not executing a submitted task constitutes absence of liveness, at least from the perspective of the client that submitted the task.

This last example of atomic updates to a set of configuration znodes is different: it is about safety, not liveness. Reading the znodes while they are being updated might lead to a client reading an inconsistent configuration. The `invalid` znode makes sure that clients read the state only when a valid configuration is available.

For the liveness examples we have seen, the order of delivery of notifications is not particularly important. As long as clients eventually learn of those events, they will make progress. For safety, however, receiving a notification out of order might lead to incorrect behavior.

## The Herd Effect and the Scalability of Watches

One issue to be aware of is that ZooKeeper triggers all watches set for a particular znode change when the change occurs. If there are 1,000 clients that have set a watch on a given znode with a call to `exists`, then 1,000 notifications will be sent out when the znode is created. A change to a watched znode might consequently generate a spike of notifications. Such a spike could affect, for example, the latency of operations submitted around the time of the spike. When possible, we recommend avoiding such a use of ZooKeeper in which a large number of clients watch for a change to a given znode. It is much better to have only a few clients watching any given znode at a time, and ideally at most one.

One way around this problem that doesn't apply in every case but might be useful in some is the following. Say that *n* clients are competing to acquire a lock (e.g., a master lock). To acquire the lock, a process simply tries to create the `/lock` znode. If the znode exists, the client watches the znode for deletion. When the znode is deleted, the client tries again to create `/lock`. With this strategy, all clients watching `/lock` receive a notification when `/lock` is deleted. A different approach is to have each client create a sequential znode `/lock/lock-`. Recall that ZooKeeper adds a sequence number to the znode, automatically making it `/lock/lock-`*`xxx`*, where *`xxx`* is a sequence number. We can use the sequence number to determine which client acquires the lock by granting it to the client that created the znode under `/lock` with the smallest sequence number. In this scheme, a client determines if it has the smallest sequence number by getting the children of `/lock` with `getChildren`. If the client doesn't have the smallest sequence number, it watches the next znode in the sequence determined by the sequence numbers. For example, say we have three znodes: `/lock/lock-001`, `/lock/lock-002`, and `/lock/lock-003`. In this example:

- The client that created `/lock/lock-001` has the lock.
- The client that created `/lock/lock-002` watches `/lock/lock-001`.
- The client that created `/lock/lock-003` watches `/lock/lock-002`.

This way each node has at most one client watching it.

Another dimension to be aware of is the state generated with watches on the server side. Setting a watch creates a `Watcher` object on the server. According to the YourKit profiler, setting a watch adds around 250 to 300 bytes to the amount of memory consumed by the watch manager of a server. Having a very large number of watches implies that the watch manager consumes a nonnegligible amount of server memory. For example, having 1 million outstanding watches gives us a ballpark figure of 0.3 GB. Consequently, a developer must be mindful of the number of watches outstanding at any time.

## Takeaway Messages

In a distributed system, there are many events that trigger actions. ZooKeeper provides efficient mechanisms for keeping track of important events that require processes in the system to react. Examples we have discussed here are related to the regular flow of applications (e.g., execution of tasks) or crash faults (e.g., master crashes).

One key ZooKeeper feature that we have used is *notifications*. ZooKeeper clients register watches with ZooKeeper to receive notifications upon changes to the ZooKeeper state. The order of notifications delivered is important; clients must not observe different orders for the changes to the ZooKeeper state.

One particular feature that is useful when dealing with changes is the `multi` call. It enables multiple operations to be executed in a block and often avoids race conditions in distributed applications when clients are reacting to events and changing the ZooKeeper state.

We expect most applications to follow the pattern we present here, although variants are of course possible and acceptable. We have focused on the asynchronous API because we encourage developers to use it. The asynchronous API enables applications to use ZooKeeper resources more efficiently and to obtain higher performance.

# CHAPTER 5
# Dealing with Failure

Life would be so much easier if failures never happened. Of course, without failures, much of the need for ZooKeeper would also go away. To effectively use ZooKeeper it is important to understand the kinds of failures that happen and how to handle them.

There are three main places where failures occur: in the ZooKeeper service itself, the network, and an application process. Recovery depends on finding which one of these is the locus of the failure, but unfortunately, doing so isn't always easy.

Imagine the simple configuration shown in Figure 5-1. Just two processes make up the application, and three servers make up the ZooKeeper service. The processes will connect to one of the servers at random and so may end up connecting to different servers. The servers use internal protocols to keep the state in sync across clients and present a consistent view to clients.

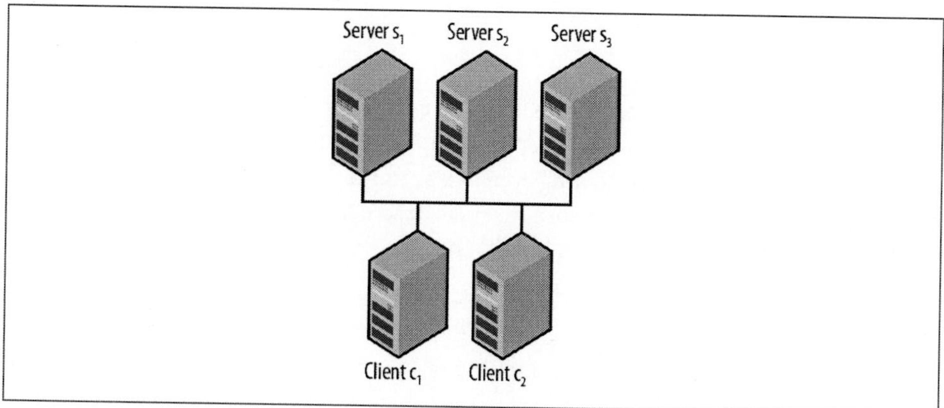


*Figure 5-1. Simple distributed application diagram*

Figure 5-2 shows some of the failures that can happen in the various components of the system. It's interesting to examine how an application can distinguish between the different types of failures. For example, if the network is down, how can $c_1$ determine the difference between a network failure and a ZooKeeper service outage? If ZooKeeper $s_1$ is the only server that is down, the other ZooKeeper servers will be online, so if there is no network problem, $c_1$ will be able to connect to one of the other servers. However, if $c_1$ cannot connect to any of the servers, it may be because the service is unavailable (perhaps because a majority of the servers are down) or because there is a network failure.

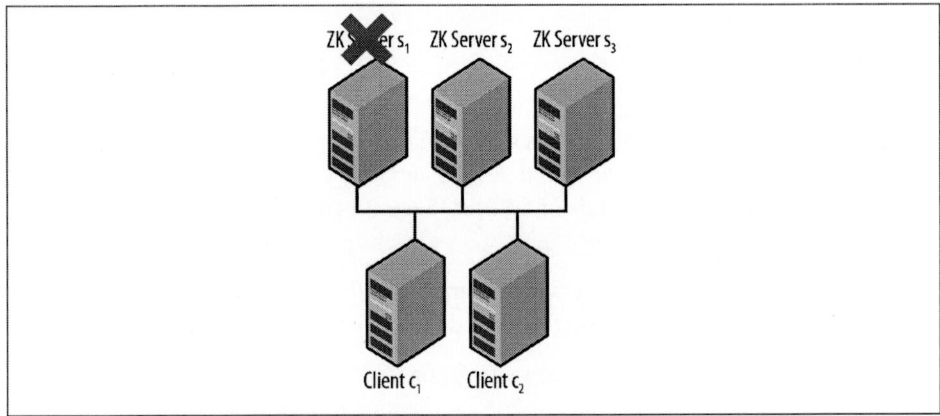


*Figure 5-2. Simple distributed application diagram with failures*

This example shows that it isn't always possible to handle failures based on the component in which they occur, so instead ZooKeeper presents its view of the system and developers work within that view.

If we examine Figure 5-2 from the perspective of $c_2$, we see that a network failure that lasts long enough will cause the session between $c_1$ and ZooKeeper to expire. Thus, even though $c_1$ is actually still alive, ZooKeeper will declare $c_1$ to be dead because $c_1$ cannot connect to any server. If $c_1$ is watching ephemeral znodes from $c_1$, it will be informed of the death of $c_1$. So $c_2$ will know for sure that $c_1$ is dead because ZooKeeper will have told it so, even though in this scenario $c_1$ is still alive.

In this scenario, $c_1$ cannot talk to the ZooKeeper service. It knows it is alive, but it cannot be sure whether or not ZooKeeper has declared it dead, so it must assume the worst. If $c_1$ takes an action that a dead process should not take (changing external resources, for example), it can corrupt the system. If $c_1$ is ever able to reconnect with ZooKeeper and finds out that its session is no longer active, it needs to make sure to stay consistent with the rest of the system and terminate or execute restart logic to appear as a new instance of the process.

**Second-Guessing ZooKeeper**

It may be tempting to second-guess ZooKeeper. It has been done before. Unfortunately, the uncertainty problem is fundamental, as we pointed out in the introduction to this chapter. The second-guesser may indeed guess correctly when ZooKeeper is wrong, but she may also guess incorrectly when ZooKeeper is right. System design is simpler and failures are easier to understand and diagnose if ZooKeeper is the designated source of truth.

Rather than trying to determine causes of failures, ZooKeeper exposes two classes of failures: *recoverable* and *unrecoverable*. Recoverable failures are transient and should be considered relatively normal—things happen. Brief network hiccups and server failures can cause these kinds of failures. Developers should write their code so that their applications keep running in spite of these failures.

Unrecoverable failures are much more problematic. These kinds of failures cause the ZooKeeper handle to become inoperable. The easiest and most common way to deal with this kind of failure is to exit the application. Examples of causes of this class of failure are session timeouts, network outages for longer than the session timeout, and authentication failures.

# Recoverable Failures

ZooKeeper presents a consistent state to all of the client processes that are using it. When a client gets a response from ZooKeeper, the client can be confident that the response will be consistent with all other responses that it or any other client receives. There are times when a ZooKeeper client library loses its connection with the ZooKeeper service and can no longer provide information that it can guarantee to be consistent. When a ZooKeeper client library finds itself in this situation, it uses the `Disconnected` event and the `ConnectionLossException` to express its lack of knowledge about the state of the system.

Of course, the ZooKeeper client library vigorously tries to extricate itself from this situation. It will continuously try to reconnect to another ZooKeeper server until it is finally able to reestablish the session. Once the session is reestablished, ZooKeeper will generate a `SyncConnected` event and start processing requests. ZooKeeper will also reregister any watches that were previously registered and generate watch events for any changes that happened during the disconnection.

A typical cause of `Disconnected` events and `ConnectionLossExceptions` is a ZooKeeper server failure. Figure 5-3 shows an example of such a failure. In this example a client is connected to server $s_2$, which is one of two active ZooKeeper servers. When $s_2$ fails, the client's `Watcher` objects will get a `Disconnected` event and any pending requests will return with a `ConnectionLossException`. The ZooKeeper service itself is fine because a majority of servers are still active, and the client will quickly reestablish its session with a new server.

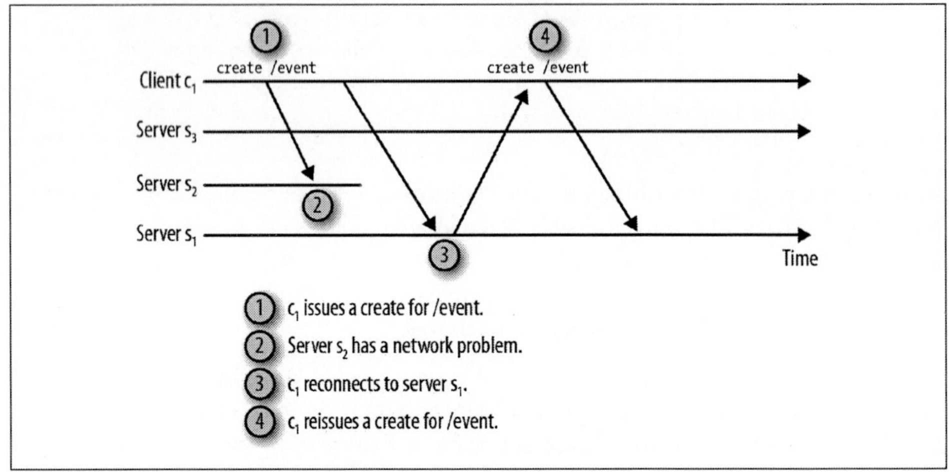


*Figure 5-3. Connection loss example*

If the client doesn't have any pending requests, all this will take place with very little disruption to the client. Apart from a `Disconnected` event followed by a `SyncConnected` event, the client will not notice the change. If there are pending requests, however, the connection loss is much more disruptive.

If the client has a pending request outstanding, such as a `create` request that it just submitted, when the connection loss happens the client will get a `ConnectionLossException` for synchronous requests and a `CONNECTIONLOSS` return code for asynchronous requests. However, the client will not be able to tell from these exceptions or return codes whether or not the requests were processed. As we have seen, handling the connection loss complicates the code because the application code must figure out whether the requests actually completed. One very bad way of dealing with the complication of handling connection loss is to code for the simple case, then shut everything down and restart if a `ConnectionLossException` or `CONNECTIONLOSS` return code is received. Although this makes the code simpler, it turns what should be a minor disruption into a major system event.

To see why, let's look at a system that is composed of 90 client processes connected to a ZooKeeper cluster of three servers. If the application is written using this simple but bad style and one of the ZooKeeper servers fails, 30 client processes will shut down and restart their sessions with ZooKeeper. To make matters worse, the session shutdown happens when the client processes are not connected to ZooKeeper, so their sessions will not get explicitly shut down and ZooKeeper will have to detect the failures based on the session timeouts. The end result is that a third of the application processes restart, and the restarts may be delayed because the new processes must wait for the locks held by the old sessions to expire. On the other hand, if the application is written to correctly handle connection loss, such scenarios will cause very little system disruption.

Developers must keep in mind that while a process is disconnected, it cannot receive updates from ZooKeeper. Though this may sound obvious, an important state change that a process may miss is the death of its session. Figure 5-4 shows an example of such a scenario. Client $c_1$, which happens to be a leader, loses its connection at time $t_1$, but it doesn't find out that it has been declared dead until time $t_4$. In the meantime, its session expires at time $t_2$, and at time $t_3$ another process becomes the leader. From time $t_2$ to time $t_4$ the old leader does not know that it has been declared dead and another leader has taken control.

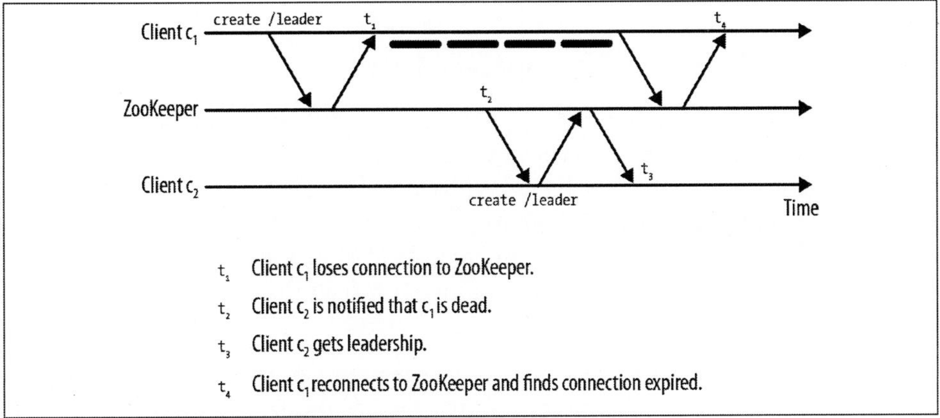


*Figure 5-4. Revenge of the living dead*

If the developer is not careful, the old leader will continue to act as a leader and may take actions that conflict with those of the new leader. For this reason, when a process receives a `Disconnected` event, the process should suspend actions taken as a leader until it reconnects. Normally this reconnect happens very quickly.

If the client is disconnected for an extended period of time, the process may choose to close the session. Of course, if the client is disconnected, closing the session will not make ZooKeeper close sooner. The ZooKeeper service still waits for the session expiration time to pass before declaring the session expired.

**Ridiculously Long Delay to Expire**

When disconnects do happen, the common case should be a very quick reconnect to another server, but an extended network outage may introduce a long delay before a client can reconnect to the ZooKeeper service. Some developers wonder why the ZooKeeper client library doesn't simply decide at some point (perhaps twice the session timeout) that enough is enough and kill the session itself.

There are two answers to this. First, ZooKeeper leaves this kind of policy decision up to the developer. Developers can easily implement such a policy by closing the handle themselves. Second, when a ZooKeeper ensemble goes down, time freezes. Thus, when the ensemble is brought back up, session timeouts are restarted. If processes using ZooKeeper hang in there, they may find out that the long timeout was due to an extended ensemble failure that has recovered and pick right up where they left off without any additional startup delay.

## The Exists Watch and the Disconnected Event

To make session disconnection and reestablishment a little more seamless, the ZooKeeper client library will reestablish any existing watches on the new server. When the client library connects to a ZooKeeper server, it will send the list of outstanding watches and the last zxid (the last state timestamp) it has seen. The server will go through the watches and check the modification timestamps of the znodes that correspond to them. If any watched znodes have a modification timestamp later than the last zxid seen, the server will trigger the watch.

This logic works perfectly for every ZooKeeper operation except `exists`. The `exists` operation is different from all other operations because it can set a watch on a znode that does not exist. If we look closely at the watch registration logic in the previous paragraph, we see that there is a corner case in which we can miss a watch event.

Figure 5-5 illustrates the corner case that causes us to miss the creation event of a watched znode. In this example, the client is watching for the creation of `/event`. However, just as the `/event` is created by another client, the watching client loses its connection to ZooKeeper. During this time the other client deletes `/event`, so when the watching client reconnects to ZooKeeper and reregisters its watch, the ZooKeeper server no longer has the `/event` znode. Thus, when it processes the registered watches and sees the watch for `/event`, and sees that there is no node called `/event`, it simply reregisters the watch, causing the client to miss the creation event for `/event`. Because of this corner case, you should try to avoid watching for the creation event of a znode. If you do watch for a creation event it should be for a long-lived znode; otherwise, this corner case can bite you.

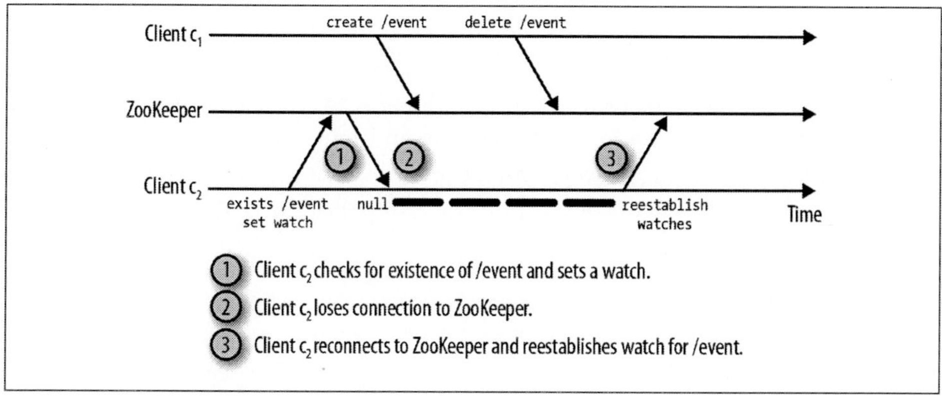


*Figure 5-5. Notification corner case*

**Hazards of Automatic Disconnect Handling**

There are ZooKeeper wrapper libraries that will automatically handle connection loss failures for you by simply reissuing commands. In some cases this is perfectly acceptable, but other cases may result in errors. For example, if the znode `/leader` is used to establish leadership and your process issues a `create` for `/leader` that results in a connection loss, a blind retry of the `create` will cause the second `create` to fail because `/leader` already exists, so the process will assume that another process has leadership. Of course, you can recognize and handle this situation if you are aware of this case and understand how the wrapper library works. Some libraries are much more sophisticated, so if you do use one of these libraries, it is good to have an understanding of ZooKeeper and a strong understanding of the guarantee that the library is providing you.

# Unrecoverable Failures

Occasionally, such bad things happen that a session cannot be recovered and must be closed. The most common reason for this is that the session expires. Another reason is that an authenticated session can no longer authenticate itself to the server. In both cases, ZooKeeper throws away the session state.

The clearest example of this lost state is the ephemeral znodes that get deleted when a session is closed. Internally, ZooKeeper keeps a less visible state that is also discarded when a session is closed.

An unrecoverable failure happens when the client fails to provide proper credentials to authenticate the session, or when it reconnects to an expired session after a `Disconnected` event. The client library does not determine that a session has failed on its own—

as we saw in Figure 5-4, the old client did not figure out that it was disconnected until time $t_4$, long after it was declared dead by the rest of the system.

The easiest way to deal with unrecoverable failures is to terminate the process and restart. This allows the process to come back up and reinitialize its state with a new session. If the process is going to keep running, it must clear out any internal application process state associated with the old session and reinitialize with a new one.

**Hazards of Automatic Recovery from Unrecoverable Failures**

It is tempting to automatically recover from unrecoverable failures by simply re-creating a new ZooKeeper handle under the covers. In fact, that is exactly what early ZooKeeper implementations did, but early users noted that this caused problems. A process that thought it was a leader could lose its session, but before it could notify its other management threads that it was no longer the leader, these threads were manipulating data using the new handle that should be accessed only by a leader. By making a one-to-one correspondence between a handle and a session, ZooKeeper now avoids this problem. There may be cases where automatic recovery is fine, especially in cases where a client is only reading data, but it is important that clients making changes to ZooKeeper data keep in mind the hazards of automatic recovery from session failures.

## Leader Election and External Resources

ZooKeeper presents a consistent view of the system to all of its clients. As long as clients do all their interactions through ZooKeeper (as our examples have done), ZooKeeper will keep everything in sync. However, interactions with external devices will not be fully protected by ZooKeeper. A particularly problematic illustration of this lack of protection, and one that has often been observed in real settings, happens with overloaded host machines.

When the host machine on which a client process runs gets overloaded, it will start swapping, thrashing, or otherwise cause large delays in processes as they compete for overcommitted host resources. This affects the timeliness of interactions with ZooKeeper. On the one hand, ZooKeeper will not be able to send heartbeats in a timely manner to ZooKeeper servers, causing ZooKeeper to time out the session. On the other hand, scheduling of local threads on the host machine can cause unpredictable scheduling: an application thread may believe a session is active and a master lock is held even though the ZooKeeper thread will signal that the session has timed out when the thread has a chance to run.

Figure 5-6 shows a problematic issue with this timeline. In this example, the application uses ZooKeeper to ensure that only one master at a time has exclusive access to an external resource. This is a common method of centralizing management of the resource to ensure consistency. At the start of the timeline, Client $c_1$ is the master and has exclusive access to the external resource. Events proceed as follows:

1. At $t_1$, $c_1$ becomes unresponsive due to overload and stops communicating with ZooKeeper. It has queued up changes to the external resource but has not yet received the CPU cycles to send them.
2. At $t_2$, ZooKeeper declares $c_1$'s session with ZooKeeper dead. At this time it also deletes all ephemeral nodes associated with $c_1$'s sessions, including the ephemeral node that it created to become the master.
3. At $t_3$, $c_2$ becomes the master.
4. At $t_4$, $c_2$ changes the state of the external resource.
5. At $t_5$, $c_1$'s overload subsides and it sends its queued changes to the external resource.
6. At $t_6$, $c_1$ is able to reconnect to ZooKeeper, finds out that its session has expired, and relinquishes mastership. Unfortunately, the damage has been done: at time $t_5$, changes were made to the external resource, resulting in corruption.

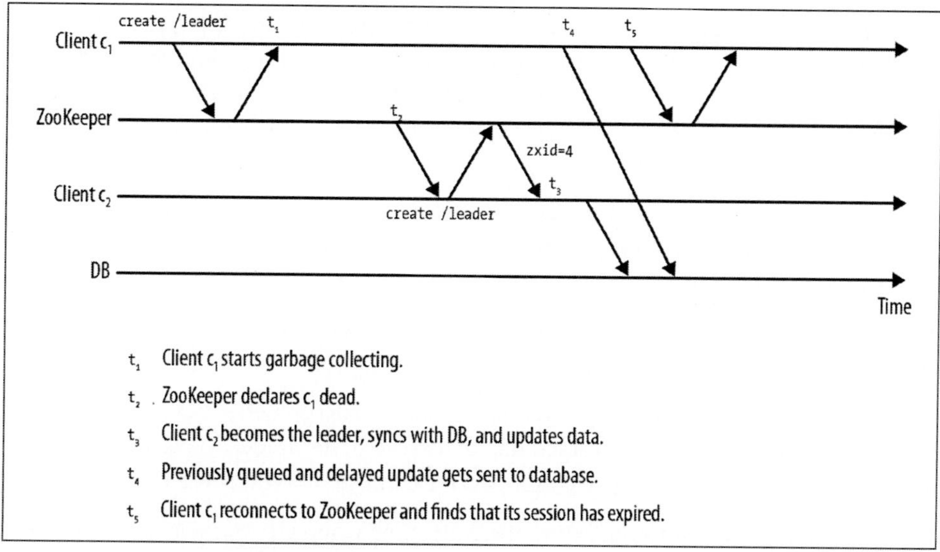


*Figure 5-6. Coordinating external resources*

Apache HBase, one of the early adopters of ZooKeeper, ran into this problem in the field. HBase has region servers that manage regions of a database table. The data is stored on a distributed file system, HDFS, and a region server has exclusive access to the files that correspond to its region. HBase ensures that only one region server is active at a time for a particular region by using leader election through ZooKeeper.

The region server is written in Java and has a large memory footprint. When available memory starts getting low, Java starts periodically running garbage collection to find memory no longer in use and free it for future allocations. Unfortunately, when collecting lots of memory, a long stop-the-world garbage collection cycle will occasionally run, pausing the process for extended periods of time. The HBase community found that sometimes this stop-the-world time period would be tens of seconds, which would cause ZooKeeper to consider the region server as dead. When the garbage collection finished and the region server continued processing, sometimes the first thing it would do would be to make changes to the distributed file system. This would end up corrupting data being managed by the new region server that had replaced the region server that had been given up for dead.

Similar problems can also result because of clock drift. In the HBase situation, time froze due to system overload. Sometimes with clock drift, time will slow down or even go backward, giving the client the impression that it is still safely within the timeout period and therefore still has mastership, even though its session has already expired with ZooKeeper.

There are a couple of approaches to addressing this problem. One approach is to make sure that you don't run into overload and clock drift situations. Careful monitoring of system load can help detect possibly problematic situations, well-designed multithreaded applications can avoid inducing overloads, and clock synchronization programs can keep system clocks in sync.

Another approach is to extend the coordination data provided by ZooKeeper to the external devices, using a technique called *fencing*. This is used often in distributed systems to ensure exclusive access to a resource.

We will show an example of implementing simple fencing using a fencing token. As long as a client holds the most recent token, it can access the resource.

When we create a leader znode, we get back a `Stat` structure. One of the members of that structure is the `czxid`, which is the zxid that created the znode. The zxid is a unique, monotonically increasing sequence number. We can use the `czxid` as a fencing token.

When we make a request to the external resource, or when we connect to the external resource, we also supply the fencing token. If the external resource has received a request or connection with a higher fencing token, our request or connection will be rejected. This means that once a new master connects to an external resource and starts managing it, if an old master tries to do anything with the external resource, its requests will fail; it will be fenced off. Fencing has the nice benefit that it will work reliably even in the presence of system overload or clock drift.

Figure 5-7 shows how this technique solves the scenario of Figure 5-6. When $c_1$ becomes the leader at time $t_1$, the creation zxid of the `/leader` znode is 3 (in reality, the zxid would be a much larger number). It supplies the creation zxid as the fencing token to connect with the database. Later, when $c_1$ becomes unresponsive due to overload, ZooKeeper declares $c_1$ as failed and $c_2$ becomes the new leader at time $t_2$. $c_2$ uses 4 as its fencing token because the `/leader` znode it created has a creation zxid of 4. At time $t_3$, $c_2$ starts making requests to the database using its fencing token. Now when $c_1$'s request arrives at the database at time $t_4$, it is rejected because its fencing token (3) is lower than the highest-seen fencing token (4), thus avoiding corruption.

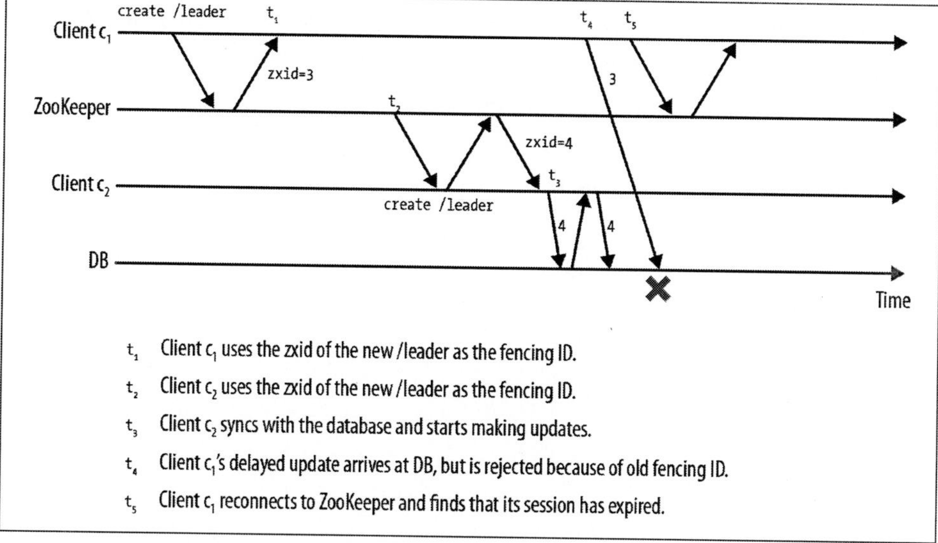

create /leader
Client c1
ZooKeeper
Client c2
DB
zxid=3
zxid=4
create /leader
Time
t1 Client c1 uses the zxid of the new /leader as the fencing ID.
t2 Client c2 uses the zxid of the new /leader as the fencing ID.
t3 Client c2 syncs with the database and starts making updates.
t4 Client c1's delayed update arrives at DB, but is rejected because of old fencing ID.
t5 Client c1 reconnects to ZooKeeper and finds that its session has expired.

*Figure 5-7. Fencing with ZooKeeper*

Unfortunately, this fencing scheme requires changes to the protocol between the client and the resource. There must be room in the protocol to add the zxid, and the external resource needs a persistent store to track the latest zxid received.

Some external resources, such as some file servers, provide locking to partially address this problem of fencing. Unfortunately, such locking also has limitations. A leader that has been swapped out and declared dead by ZooKeeper may still hold a valid lock and therefore prevent a newly elected leader from acquiring the lock it needs to make progress. In such cases, it may be more practical to use the resource lock to determine leadership and have the leader create the `/leader` znode for informational purposes.

## Takeaway Messages

Failures are a fact of life in any distributed system. ZooKeeper doesn't make the failures go away, but it does provide a framework to handle them. To handle failures effectively, developers that use ZooKeeper need to react to the state change events and failure codes and exceptions thrown by ZooKeeper. Unfortunately, not all failures are handled in the same way in all cases. Developers must consider that there are times in the disconnected state, or when dealing with disconnected exceptions, that the process does not know what is happening in the rest of the system or even whether its own pending requests have executed. During periods of disconnection, processes cannot assume that the rest of the system still believes they are running, and even though the ZooKeeper client library will reconnect with ZooKeeper servers and reestablish watches, the process must still validate the results of any pending requests that may or may not have executed.

CHAPTER 6

# ZooKeeper Caveat Emptor

The previous chapters discussed how to code with ZooKeeper, implementing some basic cases as well as a few advanced ones. In this chapter, we focus on some tricky aspects of ZooKeeper, mostly related to session semantics and ordering. The material covered here might not affect your development, but it is good to be aware of these issues in case you come across any of them.

The structure of this chapter is different from the others: we present it as a mix of issues without a linear flow. Each section is self-contained, so they can be read separately. We have discussed some tricky issues in earlier chapters, but this chapter puts together some others that it didn't make sense to discuss in other places. They are still important issues, because many developers have stumbled upon them.

## Using ACLs

Normally you would expect access control to be described in an administration section. However, in ZooKeeper the developer, rather than the administrator, is usually the one who manages access control. This is because access rights must be set every time a znode is created; it does not inherit the access permissions from its parent. Access checks are also done on a per-znode basis. If a client has access to a znode, it can access it even if that client cannot access the parent of the znode.

ZooKeeper controls access using access control lists (ACLs). An ACL contains entries of the form `scheme:auth-info`, where `scheme` corresponds to a set of built-in authentication schemes and `auth-info` encodes the authentication information in some manner specific to the scheme. ZooKeeper enforces security by checking the authorization information submitted by the client process upon access to each znode. If a process has not supplied authentication information, or if this information does not match what is needed to execute a request on a znode, the process will receive a permission error.

To add authentication information to a ZooKeeper handle, issue the `addAuthInfo` call in the format:

```
void addAuthInfo(
    String scheme,
    byte auth[]
    )
```

where:

`scheme`
: The scheme used to authenticate.

`auth`
: The authentication information to be sent to the server. The type of this parameter is `byte []`, but most of the current schemes take a `String`, so usually you will convert the `String` to a `byte []` using `String.getBytes()`.

A process can add authentication information using `addAuthInfo` at any time. Usually, it will be called right after the ZooKeeper handle is created. A process can call this method multiple times to add multiple identities to a ZooKeeper handle.

## Built-in Authentication Schemes

ZooKeeper offers four built-in schemes to handle ACLs. One of them we have been using implicitly through the `OPEN_ACL_UNSAFE` constant. That ACL uses the `world` scheme that just lists `anyone` as the `auth-info`. `anyone` is the only `auth-info` that can be used with the `world` scheme.

Another special built-in scheme used by administrators is the `super` scheme. This scheme is never included in any ACL, but it can be used to authenticate to ZooKeeper. A client that authenticates with `super` will not be restricted by ACLs of any of the znodes. "Authentication and Authorization Options" on page 186 has more information about the `super` scheme.

We will look at the other two schemes in the following example.

When ZooKeeper starts with an empty tree, there is one starting znode: `/`. This znode is open to everyone. Let's assume that Amy the administrator is setting up the ZooKeeper service. Amy also creates the `/apps` znode to be the parent of znodes created for applications that will use the service. She wants to lock the service down a bit, so she sets the ACL for `/` and `/apps` to:

```
digest:amy:Iq0onHjzb4KyxPAp8YWOIC8zzwY=, READ | WRITE | CREATE | DELETE | ADMIN
```

This ACL has exactly one entry, giving Amy all access rights. Amy has chosen `amy` as her user ID.

digest is a built-in scheme whose auth-info has the form userid:passwd_digest when setting the ACL and userid:password when calling addAuthInfo. The passwd_digest is a cryptographic digest of the user's password. In the example ACL, Iq0onHjzb4KyxPAp8YWOIC8zzwY= is the passwd_digest, so when Amy called addAuthInfo, auth would be the byte array corresponding to the string amy:secret.

Amy used DigestAuthenticationProvider as follows to generate a digest for her account amy with password secret:

```
java -cp $ZK_CLASSPATH \
    org.apache.zookeeper.server.auth.DigestAuthenticationProvider amy:secret
....
amy:secret->amy:Iq0onHjzb4KyxPAp8YWOIC8zzwY=
```

The funny string following amy: is the password digest. This is what we used for the ACL entry. When Amy authenticates to ZooKeeper, she will use digest amy:secret. For example, if Amy is using *zkCli.sh* to connect to ZooKeeper, she can authenticate using:

```
[zk: localhost:2181(CONNECTED) 1] addauth digest amy:secret
```

To avoid writing out the full digests in the following examples, we will simply use XXXXX as a placeholder for the digest.

Amy wants to set up a subtree for a new application called SuperApp that is being developed by Dom the Developer, so she creates /apps/SuperApp and sets the ACL to be:

```
digest:dom:XXXXX, READ | WRITE | CREATE | DELETE | ADMIN
digest:amy:XXXXX, READ | WRITE | CREATE | DELETE | ADMIN
```

This ACL is composed of two entries, one for Dom and one for Amy. The entries give full privileges to clients who can produce the password of either dom or amy.

Note that, according to Dom's entry in the ACL, he has ADMIN permission to /apps/SuperApp, which means that Dom can remove Amy's access to /apps/SuperApp by changing the ACL to not include her entry. Of course, Amy has access to the super secret, so she can always access any znode even if Dom removes her access to it.

Dom uses ZooKeeper to store the configuration of his application, so he creates /apps/SuperApp/config to store the configuration. He then creates the znode using the ACL that we have used in all of our examples, OPEN_ACL_UNSAFE. Dom thinks that because access is restricted to /apps and /apps/SuperApp, /apps/SuperApp/config is protected. As we will see, we don't call it UNSAFE for nothing.

Let's say there is a guest named Gabe who has network access to the ZooKeeper service. Because of the ACLs, Gabe cannot access /app or /apps/SuperApp. Gabe cannot list the children of /apps/SuperApp, for example. But perhaps Gabe has an idea that Dom uses

ZooKeeper for configuration, and `config` is a rather obvious name for a configuration file, so he connects to the ZooKeeper service and calls `getData` on `/apps/SuperApp/config`. Because of the open ACL used on that znode, Gabe can read the file. But it doesn't stop there. Gabe can change, delete, and even restrict access to `/apps/SuperApp/config`.

Let's assume that Dom realizes this and changes the ACL of `/apps/SuperApp/config` to:

```
digest:dom:XXXXX, READ | WRITE | CREATE | DELETE | ADMIN
```

As things progress, Dom gets a new developer, Nico, to help finish SuperApp. Nico needs access to the SuperApp subtree, so Dom changes the ACL of files in that subtree to include Nico. The new ACL is:

```
digest:dom:XXXXX, READ | WRITE | CREATE | DELETE | ADMIN
digest:nico:XXXXX, READ | WRITE | CREATE | DELETE | ADMIN
```

**Where Do Digest Usernames and Passwords Come From?**

You may notice that the usernames and passwords that are used with the digest scheme seem to be appearing from thin air. Actually, that is exactly where they come from. They don't need to correspond to any real system identity. Usernames can also overlap. There may be another developer named Amy that starts working with Dom and Nico. Dom can add her into ACLs using `amy:XXXXX`. The only clash would be if both Amys happened to choose the same password, because they would then have access to each other's files.

Now Dom and Nico will have the access they need to finish working on SuperApp. When the application goes into production, though, Dom and Nico may not want to give the secret passwords to the processes that need to access the ZooKeeper data. So they decide to restrict access to the data based on the network address of the machines that are running the SuperApp processes. All of the machines are on the 10.11.12.0/24 network. They therefore change the ACL of the znodes in the SuperApp subtree to:

```
digest:dom:XXXXX, READ | WRITE | CREATE | DELETE | ADMIN
digest:nico:XXXXX, READ | WRITE | CREATE | DELETE | ADMIN
ip:10.11.12.0/24, READ
```

The `ip` scheme takes the network address and mask. Because it uses the address of the client to do the ACL check, clients do not need to call `addAuthInfo` with the `ip` scheme to access a znode using this ACL.

Now any ZooKeeper client that connects from the 10.11.12.0/24 network will have read access to znodes in the SuperApp subtree. This authentication scheme assumes that IP addresses cannot be spoofed, which is a rather naive assumption that may not be appropriate for all environments.

### SASL and Kerberos

There are a couple of problems with the example in the previous section. First, if new developers join or leave the group, an administrator has to change all the ACLs; it would be nice if we could avoid this by using groups. Second, we also have to change all the ACLs if we want to change any of the passwords of any of the developers. Finally, neither the `digest` nor the `ip` scheme is appropriate if the network is not trusted. A scheme called `sasl`, which ships with ZooKeeper, addresses these issues.

SASL stands for Simple Authentication and Security Layer. It is a framework that abstracts the underlying system of authentication so that applications that use SASL can use any of the various protocols supported by SASL. With respect to ZooKeeper, SASL usually uses Kerberos, which is an authentication protocol that provides the missing features that we mentioned earlier. SASL uses `sasl` for its scheme name, and the `id` is the Kerberos ID of the client.

SASL is an extended ZooKeeper scheme. That means it needs to be enabled at runtime with a configuration parameter or Java system property. If you want to add it to the ZooKeeper configuration file, use the configuration parameter `authProvider.`*XXX*. If you want to use a system property instead, use the `zookeeper.authProvider.`*XXX* property name. In both cases, *XXX* can be anything as long as there are no other `authProviders` with the same *XXX*. Usually *XXX* is a number starting at `0`. The value of the parameter or property is `org.apache.zookeeper.server.auth.SASLAuthenticationProvider`. This will enable SASL.

### Adding New Schemes

Many other possible schemes could be used with ZooKeeper. Making them available is a "simple matter of coding." The `org.apache.zookeeper.server.auth` package offers an interface called `AuthenticationProvider`. If you implement your own provider, you can enable it by putting your new classes in the server's classpath and creating a Java system property whose key has the `zookeeper.authProvider.` prefix and whose value is the name of the class that implements `AuthenticationProvider`.

## Session Recovery

Suppose your ZooKeeper client crashes and recovers. When it comes back, there are a couple of issues that the application needs to be aware of. First, the state of ZooKeeper might not be the same as it was at the time the client crashed. As time elapses, other clients might have made progress and changed the ZooKeeper state. Consequently, it is recommended that the client not try to persist any cached state that comes from ZooKeeper, but instead uses ZooKeeper as the "source of truth" for all coordination state.

For example, in our master-worker implementation, if the primary master crashes and recovers, the ensemble might have failed over in the meantime to a backup master that has assigned tasks. When the first master recovers, it shouldn't assume that it is still the master or that the list of pending assignments hasn't changed.

The second important issue is that the client operations to ZooKeeper that were pending at the time the client crashed might have completed. Given that the client crashed before it received a confirmation, ZooKeeper can't guarantee that the operations have been executed successfully. Consequently, upon recovery, a client might need to perform some cleanup on the ZooKeeper state to complete some operations. For example, if our master crashes before deleting a pending task that has been assigned, it needs to delete this task from ZooKeeper in case our master becomes a primary again.

Although the discussion so far has focused on the case where the client crashes, all the points in this section apply also when the session has simply expired. For a session to expire, it is not necessary for the client to crash. The session may expire due to network issues or other issues, such as garbage collection pauses in Java. In the case of session expiration, the client must take into account that the ZooKeeper state might have changed and that some of its ZooKeeper requests might not have gone through.

## Version Is Reset When Znode Is Re-Created

This might sound like a naive observation, but it is necessary to remember that when a znode is deleted and re-created, its version number is reset. An application that tries to make version checks after a znode is re-created might fall into this trap.

Say that a client gets the data of a znode (e.g., `/z`), changes the data of the znode, and writes it back under the condition that the version is 1. If the znode is deleted and re-created while the client is updating the data, the version matches, but it now contains the wrong data.

Another possible scenario is for changes to a znode to occur by deleting and re-creating the znode; the znode is never changed with `setData` but is different nevertheless. In this case, checking the version gives no clue about changes to the znode. The znode may change an arbitrary number of times and its version will still be 0.

## The sync Call

If clients of an application communicate only by reading and writing to ZooKeeper, the application shouldn't worry about `sync`. `sync` exists because communication outside ZooKeeper may lead to a problem often referred to as a *hidden channel*, explained in "Ordering Guarantees" on page 91. The idea is that a client *c′* may tell another client *c* about some change to the ZooKeeper state through a direct channel (e.g., a TCP connection between *c* and *c′*), but when *c* reads the ZooKeeper state, it doesn't observe the change.

This scenario is possible because the server serving a given client might not have processed the change yet. `sync` is supposed to deal with such situations. `sync` is an asynchronous call that a client uses before a read operation. Say that a client wants to read the znode that it has heard through a direct channel has changed. The client calls `sync`, followed by `getData`:

```
...
zk.sync(path, voidCb, ctx); ❶
zk.getData(path, watcher, dataCb, ctx); ❷
...
```

❶ `sync` takes a path, a void callback implementation, and a context object.

❷ The `getData` call is the same as before.

The path in the `sync` call documents the operation it is referring to. Internally, it doesn't really matter to ZooKeeper. When the server side handles the `sync` call, it flushes the channel between the leader and the follower that is serving the client *c* that called `sync`. Flushing here means that by the time `getData` returns, it is sure to incorporate any changes that might have happened by the time *c* calls `sync`. In the case of a hidden channel, the change communicated to *c* will have happened before the call to `sync`. Consequently, when *c* receives a response from the `getData` call, it must incorporate the change communicated directly by *c′*. Note that other changes might have happened to the same znode in the meantime, so ZooKeeper only guarantees that the change communicated directly has been incorporated in the response of `getData`.

There is a caveat to the use of `sync`, which is fairly technical and deeply entwined with ZooKeeper internals. (Feel free to skip it.) Because ZooKeeper is supposed to serve reads fast and scale for read-dominated workloads, the implementation of `sync` has been simplified and it doesn't really traverse the execution pipeline as a regular update operation, like `create`, `setData`, or `delete`. It simply reaches the leader, and the leader queues a response back to the follower that sent it. There is a small chance that the leader thinks that it is the leader *l*, but doesn't have support from a quorum any longer because the quorum now supports a different leader, *l′*. In this case, the leader *l* might not have all updates that have been processed, and the `sync` call might not be able to honor its guarantee.

The ZooKeeper implementation deals with this issue by making it unlikely that a quorum of followers will abandon a leader without the leader noticing. It does so by having the leader time out for any given follower based on the `tickTime`, while a follower decides that a leader is gone by receiving a socket exception, which occurs when the

TCP connection between them drops. The leader times out a follower much sooner than a TCP connection expiration. However, the corner case for an error is there, even though it has never been observed to our knowledge.

There have been discussions on the mailing list about changing the server handling of `sync` to traverse the pipeline and eliminate this corner case altogether. Currently, the ZooKeeper implementation relies on reasonable timing assumptions, and consequently no problem is expected.

# Ordering Guarantees

Although ZooKeeper officially guarantees the order of client operations during a session, circumstances outside the control of ZooKeeper can still change the way the order appears to a client. There are a few guidelines that the developer needs to be aware of to ensure the expected behavior. We discuss three cases here.

## Order in the Presence of Connection Loss

Upon a connection loss event, ZooKeeper cancels pending requests. For synchronous calls the library throws an exception, while for asynchronous calls the library invokes the callbacks with a return code indicating a connection loss. The client library won't try to resubmit the request once it has indicated to the application that the connection has been lost, so it's up to the application to resubmit operations that have been canceled. Consequently, the application can rely on the client library to issue all the callbacks, but it cannot rely on ZooKeeper to actually carry out the operation in the event of a connection loss.

To understand what impact this can have on an application, let's consider the following sequence of events:

1. Application submits a request to execute `Op1`.
2. Client library detects a connection loss and cancels pending request to execute `Op1`.
3. Client reconnects before the session expires.
4. Application submits a request to execute operation `Op2`.
5. `Op2` is executed successfully.
6. `Op1` returns with `CONNECTIONLOSS`.
7. Application resubmits `Op1`.

In this case, the application submitted `Op1` and `Op2` in that order and got `Op2` to execute successfully before `Op1`. When the application notices the connection loss in the callback of `Op1`, it tries to submit that request again. But suppose the client does not successfully reconnect. A new call to submit `Op1` will report connection loss again, and there is a risk that the application will enter into an unbounded loop of resubmitting the `Op1` request without connecting. To get out of this cycle, the application could set a bound on the number of attempts, or close the handle if reconnecting takes too long.

In some cases, it might be important to make sure that `Op1` is executed successfully before `Op2`. If `Op2` depends on `Op1` in some way, then to avoid having `Op2` executing successfully before `Op1`, we could simply wait for a successful execution of `Op1` before submitting `Op2`. This is the approach we have taken in most of our master-worker example code, to guarantee that the requests are executed in order. In general, this approach of waiting for the result of `Op1` is safe, but it adds a performance penalty because the application needs to wait for the result of one request to submit the next one, instead of having them processed concurrently.

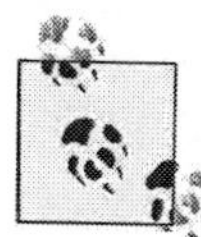

**What if We Get Rid of CONNECTIONLOSS?**

The `CONNECTIONLOSS` event exists essentially because of the case in which a request is pending when the client loses the connection. Say it is a `create` request. In such cases, the client doesn't know whether the request has gone through. The client, however, could ask the server to check whether the request has executed successfully. The server knows what has gone through, due to information cached either in memory or in its logs, so it is feasible to do it this way. If the community eventually changes ZooKeeper to access this server information when reconnecting, we will be able to remove the limitation of not being able to guarantee that a prefix executes successfully because clients will be able to reexecute pending requests when necessary. Until then, developers have to live with this limitation and deal with connection loss events.

## Order with the Synchronous API and Multiple Threads

Multithreaded applications are common these days. If you are using the synchronous API with multiple threads, it is important to pay attention to one ordering issue. A synchronous ZooKeeper call blocks until it gets a response. If two or more threads submit synchronous operations to ZooKeeper concurrently, they will each block until they receive a response. ZooKeeper will deliver the responses in order, but it is possible that due to thread scheduling, the result of an operation submitted later will be processed first. If ZooKeeper delivers responses to operations very close to each other in time, you may observe such a scenario.

If different threads submit operations concurrently, odds are that the operations are not directly related and can be executed in any order without causing consistency issues. But if the operations are related, the application client must take the order of submission into consideration when processing the results.

## Order When Mixing Synchronous and Asynchronous Calls

Here's another situation where results may appear to be out of order. Say that you submit two asynchronous operations, `Aop1` and `Aop2`. It doesn't matter what the operations are, only that they are asynchronous. In the callback of `Aop1`, you make a synchronous call, `Sop1`. The synchronous call blocks the dispatch thread of the ZooKeeper client, which causes the application client to receive the result of `Sop1` before receiving the result of `Aop2`. Consequently, the application observes the results of `Aop1`, `Sop1`, and `Aop2` in that order, which is different from the submission order.

In general, it is not a good idea to mix synchronous and asynchronous calls. There are a few exceptions—for example, when starting up, you may want to make sure that some data is in ZooKeeper before proceeding. It is possible to use Java latches and other such mechanisms in these cases, but one or more synchronous calls might also do the job.

# Data and Child Limits

ZooKeeper limits the amount of data transferred for each request to 1MB by default. This limit bounds the maximum amount of data for any given node and the number of children any parent znode can have. The choice of 1MB is somewhat arbitrary, in the sense that there is nothing fundamental in ZooKeeper that prevents it from using a different value, larger or smaller. A limit exists, however, to keep performance high. A znode with very large data, for instance, takes a long time to traverse, essentially stalling the processing pipeline while requests are being executed. The same problem happens if a client executes `getChildren` on a znode with a large number of children.

The limits ZooKeeper imposes right now for data size and number of children are already quite high, so you should avoid even getting close to these limits. We have enabled large limits to satisfy the requirements of a broader set of applications. If you have a special use case and really need different limits, you can change them as described in "Unsafe Options" on page 186.

# Embedding the ZooKeeper Server

Many developers have considered embedding the ZooKeeper server in their applications to hide their dependency on ZooKeeper. By "embedding," we mean instantiating a ZooKeeper server inside an application. The idea is to make it transparent to the application user that ZooKeeper is in use. Although the idea sounds really appealing (who likes extra dependencies, after all?), it is not really recommended. One issue we have observed with embedding is that if anything goes wrong with ZooKeeper, the user will start seeing log messages related to ZooKeeper. At that point, its use is not transparent any longer, and the application developer might not be the right expert to deal with these problems. Even worse, the availability of the application and ZooKeeper are now coupled: if one exits, the other exits too. ZooKeeper is often used to provide high availability, but embedding it in an application effectively eliminates one of its strongest benefits.

Although we do not recommend embedding ZooKeeper, there is nothing really fundamental that prevents one from doing it. ZooKeeper tests, for example, do it. Consequently, if you really think you need to follow this path, the ZooKeeper tests are a good source of ideas for how to do so.

## Takeaway Messages

Programming with ZooKeeper can be tricky at times. At a high level, we call the attention of the reader to order guarantees and the session semantics. They may seem easy to understand at first, and in a sense they are, but there are important corner cases that developers need to be aware of. Our goal in this chapter was to provide some guidelines and cover some of these corner cases.

CHAPTER 7

# The C Client

Although the Java interface to ZooKeeper is the predominant one, the C ZooKeeper client binding is also popular among ZooKeeper developers and forms the foundation for bindings in other languages. This chapter focuses on this binding. To illustrate the development of ZooKeeper applications with the C API, we'll reimplement the master of our master-worker example in C. The general idea is to expose the differences when compared to the Java API through an example.

The main reference for the C API is the *zookeeper.h* file in the ZooKeeper distribution, and the instructions to build the client library are given in the *README* file of the project distribution. Alternatively, you can use *ant compile-native*, which automates it all. Before going into code snippets, we'll give a quick summary of how to set up the development environment to help you get started.

When we build the C client, it produces two libraries: one for multithreaded clients and the other for single-threaded clients. Most of this chapter assumes the multithreaded library is being used; we discuss the single-threaded version toward the end of the chapter, but we encourage the reader to focus on multithreaded implementations.

## Setting Up the Development Environment

In the ZooKeeper distribution, we can ship precompiled JAR files that are ready to run on any platform. To compile natively using C, we need to build the required shared libraries before we can build our native C ZooKeeper applications. Fortunately, ZooKeeper has an easy way to build these libraries.

The easiest option to build the ZooKeeper native libraries is to use the *ant* build tool. In the directory where you unpacked the ZooKeeper distribution, there is a file called *build.xml*. This file has the instructions required for *ant* to build everything. You will also need *automake*, *autoconf*, and *cppunit*. These should be available in your host distribution if you are using Linux. Cygwin supplies the packages on Windows. On Mac OS X, you can use an open-source package manager such as Fink, Brew, or MacPorts.

Once all the needed applications are installed, you can build the ZooKeeper libraries using:

```
ant compile-native
```

Once the build finishes you will find libraries that you need to link with in *build/c/build/usr/lib* and include files you need in *build/c/build/usr/include/zookeeper*.

## Starting a Session

To do anything with ZooKeeper, we first need a `zhandle_t` handle. To get a handle, we call `zookeeper_init`, which has the following signature:

```
ZOOAPI zhandle_t *zookeeper_init(const char *host, ❶
                                 watcher_fn fn, ❷
                                 int recv_timeout, ❸
                                 const clientid_t *clientid, ❹
                                 void *context, ❺
                                 int flags); ❻
```

❶ String containing the host addresses of the ZooKeeper servers in the ensemble. The addresses are *`host:port`* pairs, and the pairs are comma-separated.

❷ Watcher function for processing events, defined next in this section.

❸ Session expiration time in milliseconds.

❹ Client ID of a session that has been previously established and that this client is trying to reconnect to. To obtain the client ID of an established session, we call `zoo_client_id`. Specify `0` to start a new session.

❺ Context object that is used with the returned `zkhandle_t` handle.

❻ There is no current use for this parameter, so it should be set to `0`.

**The ZOOAPI Definition**

`ZOOAPI` is used to build ZooKeeper on Windows. The possible values for `ZOOAPI` are `__declspec(dllexport)`, `__declspec(dllimport)`, and empty. The keywords `__declspec(dllexport)` and `\_\_declspec(dllimport)` export and import symbols to and from a DLL, respectively. If you are not building on Windows, leave `ZOOAPI` empty. In principle, nothing needs to be configured if you are building on Windows; the distribution configuration should be sufficient.

The call to zookeeper_init may return before the session establishment actually completes. Consequently, the session shouldn't be considered established until a ZOO_CONNECTED_STATE event has been received. This event is processed with an implementation of the watcher function, which has the following signature:

```
typedef void (*watcher_fn)(zhandle_t *zh, ❶
                           int type, ❷
                           int state, ❸
                           const char *path, ❹
                           void *watcherCtx); ❺
```

❶ ZooKeeper handle that this call to the watcher function refers to.

❷ Type of event: ZOO_CREATED_EVENT, ZOO_DELETED_EVENT, ZOO_CHANGED_EVENT, ZOO_CHILD_EVENT, or ZOO_SESSION_EVENT.

❸ State of the connection.

❹ Znode path for which the watch is triggered. If the event is a session event, the path is null.

❺ Context object for the watcher.

Here is an example of how to implement a watcher:

```
static int connected = 0;
static int expired = 0;

void main_watcher (zhandle_t *zkh,
                   int type,
                   int state,
                   const char *path,
                   void* context)
{
    if (type == ZOO_SESSION_EVENT) {
        if (state == ZOO_CONNECTED_STATE) {
            connected = 1; ❶
        } else if (state == ZOO_NOTCONNECTED_STATE ) {
            connected = 0;
        } else if (state == ZOO_EXPIRED_SESSION_STATE) {
            expired = 1; ❷
            connected = 0;
            zookeeper_close(zkh);
        }
    }
}
```

❶ Set connected upon receiving a ZOO_CONNECTED_STATE event.

❷ Set expired (and close the session handle) upon receiving a ZOO_EXPIRED_SESSION_STATE event.

**Watch Data Structures**

ZooKeeper does not have a way to remove watches so as long as watches are outstanding. Consequently, it's important to keep watch data structures around even if the process no longer cares about the session, because the completion functions may still get invoked. Java takes care of this automatically through garbage collection.

To put everything together, this is the `init` function we have for our master:

```
static int server_id;

int init (char* hostPort) {
    srand(time(NULL));
    server_id  = rand(); ❶

    zoo_set_debug_level(ZOO_LOG_LEVEL_INFO); ❷

    zh = zookeeper_init(hostPort, ❸
                        main_watcher,
                        15000,
                        0,
                        0,
                        0);

    return errno;
}
```

❶ Sets the server ID.

❷ Sets the log severity level to output.

❸ Call to create a session.

The first two lines set the seed for random number generation and set the identifier of this master. We use the `server_id` to identify different masters. (Recall that we can have one or more backup masters as well as a primary master.) Next, we set the severity level of the log messages. The implementation of logging is homebrewed (see *log.h*), and we have copied from the ZooKeeper distribution (*zookeeper_log.h*) for convenience. Finally, we have the call to `zookeeper_init`, which makes `main_watcher` the function that processes session events.

## Bootstrapping the Master

Bootstrapping the master refers to creating a few znodes used in the operation of the master-worker example and running for primary master. We first create four necessary znodes:

```
void bootstrap() {
    if(!connected) { ❶
        LOG_WARN(("Client not connected to ZooKeeper"));
        return;
    }

    create_parent("/workers", ""); ❷
    create_parent("/assign", "");
    create_parent("/tasks", "");
    create_parent("/status", "");
    ...
}
```

❶ If not yet connected, log that fact and return.

❷ Create four parent znodes: `/workers`, `/assign`, `/tasks`, and `/status`.

And here's the corresponding `create_parent` function:

```
void create_parent(const char * path,
                   const char * value) {
    zoo_acreate(zh, ❶
                path, ❷
                value, ❸
                0, ❹
                &ZOO_OPEN_ACL_UNSAFE, ❺
                0, ❻
                create_parent_completion, ❼
                NULL); ❽

}
```

❶ Asynchronous call to create a znode. It passes a `zhandle_t` instance, which is a global static variable in our implementation.

❷ The path is a parameter of the call of type `const char*`. The path is used to tie a client to a subtree of a znode, as described in "Managing Client Connect Strings" on page 197.

❸ The second parameter of the call is the data to store with the znode. We pass this data to `create_parent` just to illustrate that we need to pass it as the completion data of `zoo_create` in case we need to retry the operation. In our example, passing data to `create_parent` is not strictly necessary because it is empty in all four cases.

❹ This parameter is the length of the value being stored (the previous parameter). In this case, we set it to zero.

❺ We don't care about ACLs in this example, so we just set it to be unsafe.

❻ These parent znodes are persistent and not sequential, so we don't pass any flags.

❼ Because this is an asynchronous call, we pass a completion function that the ZooKeeper client calls upon completion of the operation.

❽ The last parameter is the context of this call, but in this particular case, there is no context to be passed.

Because this is an asynchronous call, we pass a completion function to be called when the operation completes. The definition of the completion function is:

```
typedef void
        (*string_completion_t)(int rc, ❶
                               const char *value, ❷
                               const void *data); ❸
```

❶ `rc` is the return code, which appears in all completion functions.

❷ `value` is the string returned.

❸ `data` is context data passed by the caller when making an asynchronous call. Note that the programmer is responsible for freeing any heap space associated with the data pointer.

For this particular example, we have this implementation:

```
void create_parent_completion (int rc, const char *value, const void *data) {
    switch (rc) { ❶
        case ZCONNECTIONLOSS:
            create_parent(value, (const char *) data); ❷

            break;

        case ZOK:
            LOG_INFO(("Created parent node", value));

            break;

        case ZNODEEXISTS:
            LOG_WARN(("Node already exists"));

            break;

        default:
            LOG_ERROR(("Something went wrong when running for master"));

            break;
    }
}
```

❶ Check the return code to determine what to do.

❷ Try again in the case of connection loss.

Most of the completion function consists simply of logging to inform us of what is going on. In general, completion functions are a bit more complex, although it is good practice to split functionality across different completion methods as we do in this example. Note that if a connection is lost, this code ends up calling `create_parent` multiple times. This is not a recursive call because the completion function is not called by `create_parent`. Also, `create_parent` simply calls a ZooKeeper function, so it has no side effects that would come, for example, from allocating memory space. If we do create side effects, it is important to clean up before making another call from the completion function.

The next task is to run for master. Running for master basically involves trying to create the `/master` znode to lock in the primary master role. There are a few differences from the asynchronous `create` call we just discussed for parent znodes, though:

```
void run_for_master() {
    if(!connected) {
        LOG_WARN(LOGCALLBACK(zh),
                 "Client not connected to ZooKeeper");
        return;
    }

    char server_id_string[9];
    snprintf(server_id_string, 9, "%x", server_id);
    zoo_acreate(zh,
                "/master",
                (const char *) server_id_string, ❶
                sizeof(int), ❷
                &ZOO_OPEN_ACL_UNSAFE,
                ZOO_EPHEMERAL, ❸
                master_create_completion,
                NULL);
}
```

❶ Store the server identifier in the `/master` znode.

❷ We have to pass the length of the data being stored. It is an `int`, as we have declared here.

❸ This znode is ephemeral, so we have to pass the ephemeral flag.

The completion function also has to do a bit more than the earlier one:

```
void master_create_completion (int rc, const char *value, const void *data) {
    switch (rc) {
        case ZCONNECTIONLOSS:
            check_master(); ❶

            break;

        case ZOK:
            take_leadership(); ❷
```

```
            break;

        case ZNODEEXISTS:
            master_exists(); ❸

            break;

        default:
            LOG_ERROR(LOGCALLBACK(zh),
                      "Something went wrong when running for master.");

            break;
    }
}
```

❶ Upon connection loss, check whether a master znode has been created by this master or some other master.

❷ If we have been able to create it, then take leadership.

❸ If the master znode already exists (someone else has taken the lock), run a function to watch for the later disappearance of the znode.

If this master finds that `/master` already exists, it proceeds to set a watch with a call to `zoo_awexists`:

```
void master_exists() {
    zoo_awexists(zh,
                 "/master",
                 master_exists_watcher, ❶
                 NULL,
                 master_exists_completion, ❷
                 NULL);
}
```

❶ Defines the watcher for `/master`.

❷ Callback for this `exists` call.

Note that this call allows us to pass a context to the watcher function as well. Although we do not make use of it in this case, the watcher function allows us to pass a `(void *)` to some structure or variable that represents the context of this call.

Our implementation of the watcher function that processes the notification when the znode is deleted is the following:

```
void master_exists_watcher (zhandle_t *zh,
                            int type,
                            int state,
                            const char *path,
                            void *watcherCtx) {
    if( type == ZOO_DELETED_EVENT) {
```

```
            assert( !strcmp(path, "/master") );
            run_for_master(); ❶
        } else {
            LOG_DEBUG(LOGCALLBACK(zh),
                      "Watched event: ", type2string(type));
        }
    }
```

❶ If `/master` gets deleted, run for master.

Back to the `master_exists` call. The completion function we implement is simple and follows the pattern we have been using thus far. The one small important detail to note is that between the execution of the call to create `/master` and the execution of the `exists` request, it is possible that the `/master` znode has been deleted (i.e., that the previous primary master has gone away). Consequently, the completion function verifies that the znode exists and, if it does not, the client runs for master again:

```
void master_exists_completion (int rc,
                               const struct Stat *stat,
                               const void *data) {
    switch (rc) {
        case ZCONNECTIONLOSS:
        case ZOPERATIONTIMEOUT:
            master_exists();

            break;

        case ZOK:
            if(stat == NULL) { ❶
                LOG_INFO(LOGCALLBACK(zh),
                         "Previous master is gone, running for master");
                run_for_master(); ❷
            }

            break;

        default:
            LOG_WARN(LOGCALLBACK(zh),
                     "Something went wrong when executing exists: ",
                     rc2string(rc));

            break;
    }
}
```

❶ Checks whether the znode exists by checking whether `stat` is null.

❷ Runs for master again if the znode is gone.

Once the master determines it is the primary, it takes leadership, as we explain next.

## Taking Leadership

Once the master is elected primary, it starts exercising its role. It first gets the list of available workers:

```
void take_leadership() {
    get_workers();
}

void get_workers() {
    zoo_awget_children(zh,
                       "/workers",
                       workers_watcher, ❶
                       NULL,
                       workers_completion, ❷
                       NULL);
}
```

❶ Sets a watch to run in case the list of workers changes.

❷ Defines the completion function to be called upon return.

Our implementation caches the list of workers it read last. Upon reading a new list, it replaces the old list. All this happens in the completion function of the `zoo_awget_children` call:

```
void workers_completion (int rc,
                         const struct String_vector *strings,
                         const void *data) {
    switch (rc) {
        case ZCONNECTIONLOSS:
        case ZOPERATIONTIMEOUT:
            get_workers();

            break;

        case ZOK:
            struct String_vector *tmp_workers =
                               removed_and_set(strings, &workers); ❶
            free_vector(tmp_workers); ❷
            get_tasks(); ❸

            break;

        default:
            LOG_ERROR(LOGCALLBACK(zh),
                      "Something went wrong when checking workers: %s",
                      rc2string(rc));

            break;
    }
}
```

❶ Updates the list of workers.

❷ We are not really using the list of workers that have been removed in this example, so just free it. The idea is to use it for reassignments, though. Consider doing this as an exercise.

❸ The next step is getting tasks to be assigned.

To get tasks, the server gets the children of `/tasks` and takes the ones that have been introduced since the last time the list was read. We need to take the difference because otherwise the master might end up assigning the same task twice (assigning twice is possible if we take the list of tasks as is because two consecutive reads of the children of `/tasks` might return some duplicate elements—for example, if the master does not have enough time to process all elements of the first read):

```
void get_tasks () {
    zoo_awget_children(zh,
                       "/tasks",
                       tasks_watcher,
                       NULL,
                       tasks_completion,
                       NULL);
}

void tasks_watcher (zhandle_t *zh,
                    int type,
                    int state,
                    const char *path,
                    void *watcherCtx) {
    if( type == ZOO_CHILD_EVENT) {
        assert( !strcmp(path, "/tasks") );

        get_tasks(); ❶
    } else {
        LOG_INFO(LOGCALLBACK(zh),
                 "Watched event: ",
                 type2string(type));
    }
}

void tasks_completion (int rc,
                       const struct String_vector *strings,
                       const void *data) {
    switch (rc) {
        case ZCONNECTIONLOSS:
        case ZOPERATIONTIMEOUT:
            get_tasks();

            break;

        case ZOK:
```

```
            LOG_DEBUG(LOGCALLBACK(zh), "Assigning tasks");

            struct String_vector *tmp_tasks = added_and_set(strings, &tasks);
            assign_tasks(tmp_tasks); ❷
            free_vector(tmp_tasks);

            break;
        default:
            LOG_ERROR(LOGCALLBACK(zh),
                      "Something went wrong when checking tasks: %s",
                      rc2string(rc));

            break;
    }
}
```

❶ If the list of tasks changes, get the tasks again.

❷ Assign only the tasks that are not being assigned already.

## Assigning Tasks

Assigning a task consists of getting the task data, choosing a worker, assigning the task by adding a znode to the list of tasks of the worker, and finally deleting the task from the `/tasks` znode. These actions are essentially what the following code snippets implement. Getting the task data, assigning the task, and deleting the task are all asynchronous operations and require completion functions. We begin as follows:

```
void assign_tasks(const struct String_vector *strings) {
    int i;
    for( i = 0; i < strings->count; i++) {
        get_task_data( strings->data[i] ); ❶
    }
}

void get_task_data(const char *task) {
    if(task == NULL) return;

    char * tmp_task = strndup(task, 15);
    char * path = make_path(2, "/tasks/", tmp_task);

    zoo_aget(zh,
             path,
             0,
             get_task_data_completion,
             (const void *) tmp_task); ❷
    free(path);
}

struct task_info { ❸
    char* name;
```

```
    char *value;
    int value_len;
    char* worker;
};


void get_task_data_completion(int rc, const char *value, int value_len,
                              const struct Stat *stat, const void *data) {
    int worker_index;

    switch (rc) {
        case ZCONNECTIONLOSS:
        case ZOPERATIONTIMEOUT:
            get_task_data((const char *) data);

            break;

        case ZOK:
            if(workers != NULL) {
                worker_index = (rand() % workers->count); ❹
                struct task_info *new_task; ❺
                new_task = (struct task_info*) malloc(sizeof(struct task_info));

                new_task->name = (char *) data;
                new_task->value = strndup(value, value_len);
                new_task->value_len = value_len;

                const char * worker_string = workers->data[worker_index];
                new_task->worker = strdup(worker_string);

                task_assignment(new_task); ❻
            }

            break;

        default:
            LOG_ERROR(LOGCALLBACK(zh),
                      "Something went wrong when checking the master lock: %s",
                      rc2string(rc));

            break;

    }
}
```

❶ For each task, we first get its data.

❷ Asynchronous call to get task data.

❸ Structure to keep the task context.

❹ Choose worker at random to assign the task to.

❺ Create a new `task_info` instance to store the task data.

❻ Got task data, so let's complete the assignment.

So far, the code has read the data of the task and selected a worker. The next step is to create the znode to represent the assignment:

```
void task_assignment(struct task_info *task) {
    char* path = make_path(4, "/assign/" , task->worker, "/", task->name);
    zoo_acreate(zh,
                path,
                task->value,
                task->value_len,
                &ZOO_OPEN_ACL_UNSAFE,
                0,
                task_assignment_completion,
                (const void*) task); ❶
    free(path);
}

void task_assignment_completion (int rc, const char *value, const void *data) {
    switch (rc) {
        case ZCONNECTIONLOSS:
        case ZOPERATIONTIMEOUT:
            task_assignment((struct task_info*) data);

            break;

        case ZOK:
            if(data != NULL) {
                char * del_path = "";
                del_path = make_path(2, "/tasks/",
                                     ((struct task_info*) data)->name);
                if(del_path != NULL) {
                    delete_pending_task(del_path); ❷
                }
                free(del_path);
                free_task_info((struct task_info*) data); ❸
            }

            break;

        case ZNODEEXISTS:
            LOG_DEBUG(LOGCALLBACK(zh),
                      "Assignment has alreasy been created: %s",
                      value);

            break;

        default:

            LOG_ERROR(LOGCALLBACK(zh),
                      "Something went wrong when checking the master lock: %s",
                      rc2string(rc));
```

```
                break;
            }
        }
```

❶ Create the znode representing the task assignment.

❷ Once the task has been assigned, the master deletes the task from the list of unassigned tasks.

❸ We have allocated space in the heap for the task_info instance, so we can now free it.

The final step is to delete the task from /tasks. Recall that the tasks that haven't been assigned are kept in /tasks:

```
void delete_pending_task (const char * path) {
    if(path == NULL) return;

    char * tmp_path = strdup(path);
    zoo_adelete(zh,
                tmp_path,
                -1,
                delete_task_completion,
                (const void*) tmp_path); ❶
}


void delete_task_completion(int rc, const void *data) {
    switch (rc) {
        case ZCONNECTIONLOSS:
        case ZOPERATIONTIMEOUT:
            delete_pending_task((const char *) data);

            break;

        case ZOK:
            free((char *) data); ❷
            break;

        default:
            LOG_ERROR(LOGCALLBACK(zh),
                      "Something went wrong when deleting task: %s",
                      rc2string(rc));

            break;
    }
}
```

❶ Asynchronously delete the task.

❷ There isn't much to do after the task is successfully deleted. Here, we just free the space we allocated previously to the path string.

## Single-Threaded versus Multithreaded Clients

The ZooKeeper distribution has two options for the C binding: multithreaded and single-threaded. The multithreaded version is the version we encourage developers to use, whereas the single-threaded version exists mainly for historical reasons. Back at Yahoo!, there were applications that used to run on BSD and were single-threaded. They needed a single-threaded version of the client library to be able to use ZooKeeper. If you're not in a situation that forces the use of the single-threaded library, then just go with the multithreaded library.

To use the single-threaded version, we can reuse the code shown throughout this chapter, but we need to additionally implement an event loop. For our example, it looks like this:

```
int initialized = 0;
int run = 0;
fd_set rfds, wfds, efds;

FD_ZERO(&rfds);
FD_ZERO(&wfds);
FD_ZERO(&efds);
while (!is_expired()) {
    int fd;
    int interest;
    int events;
    struct timeval tv;
    int rc;

    zookeeper_interest(zh, &fd, &interest, &tv); ❶
    if (fd != -1) {
        if (interest&ZOOKEEPER_READ) { ❷
            FD_SET(fd, &rfds);
        } else {
            FD_CLR(fd, &rfds);
        }
        if (interest&ZOOKEEPER_WRITE) { ❸
            FD_SET(fd, &wfds);
        } else {
            FD_CLR(fd, &wfds);
        }
    } else {
        fd = 0;
    }

    /*
     * Master call to get a ZooKeeper handle.
```

```
    */
    if(!initialized) {
        if(init(argv[1])) { ❹
            LOG_ERROR(("Error while initializing the master: ", errno));
        }
        initialized = 1;

    }

    /*
     * The next if block contains
     * calls to bootstrap the master
     * and run for master. We only
     * get into it when the client
     * has established a session and
     * is_connected is true.
     */
    if(is_connected() && !run) { ❺
        LOG_INFO(("Connected, going to bootstrap and run for master"));

        /*
         * Create parent znodes
         */
        bootstrap();

        /*
         * Run for master
         */
        run_for_master();

        run =1;
    }

    rc = select(fd+1, &rfds, &wfds, &efds, &tv); ❻
    events = 0;
    if (rc > 0) {
        if (FD_ISSET(fd, &rfds)) {
            events |= ZOOKEEPER_READ; ❼
        }
        if (FD_ISSET(fd, &wfds)) {
            events |= ZOOKEEPER_WRITE; ❽
        }
    }

    zookeeper_process(zh, events); ❾
}
```

❶ Return the events this client is interested in.

❷ Add `ZOOKEEPER_READ` events to the set of interests.

❸ Add `ZOOKEEPER_WRITE` events to the set of interests.

❹ Here we start application calls. This is the same `init` call shown in "Starting a Session" on page 122 to get a ZooKeeper handle.

❺ This runs after the master connects to a ZooKeeper server and receives a `ZOO_CONNECTED_EVENT`. This block bootstraps and runs for master.

❻ Use the `select` call to wait for new events.

❼ Indicates that a read event on the file descriptor `fd` has happened.

❽ Indicates that a write event on the file descriptor `fd` has happened.

❾ Processes pending ZooKeeper events. `zookeeper_process` is the same call that the multithreaded library uses to process watch events and completions. In the single-threaded case, we have to do it ourselves.

This event loop takes care of the relevant ZooKeeper events, such as callbacks and session events.

To use the multithreaded version, compile the client application with `-l zookeeper_mt` and define `THREADED` with the `-DTHREADED` option. The code uses the `THREADED` directive to indicate the part of the code in the main call that is to be executed when it is compiled with the multithreaded library. To use the single-threaded version, compile the program with `-l zookeeper_st` and without the `-DTHREADED` option. Generate the library you want to use by following the compilation procedure detailed in the ZooKeeper distribution.

**Blocking Callbacks**

If a callback blocks the thread (while it is doing disk I/O, for example), the sessions may time out because the ZooKeeper processing loops do not get the CPU time they need to do processing. This same problem does not occur with the multithreaded library because it uses separate threads for handling I/O and for completion calls.

# Takeaway Messages

The C ZooKeeper binding is very popular, and in this chapter we have explored how to develop a ZooKeeper application with it. The flow of the application is not very different from what we have already seen for the Java binding, and the key differences stem mainly from the differences between the languages. For example, here we had to deal with heap management, whereas in Java we pretty much delegate it to the JVM. We also pointed out that ZooKeeper provides the option of implementing multithreaded and single-threaded applications. We strongly encourage developers to go with the multithreaded option, but we do show how to make it single-threaded because this option comes with the distribution.

CHAPTER 8

# Curator: A High-Level API for ZooKeeper

At a high level, Curator is a set of libraries that build on top of ZooKeeper. One of the core goals of Curator is to manage the ZooKeeper handle for you, removing some (ideally all) of the complexity that connection management entails. Connection management is often tricky, as we have discussed in the past chapters, and Curator might come in handy at times.

As part of managing the handle, Curator implements a set of recipes that developers commonly use, incorporating best practices and known edge-case handling. For example, Curator implements recipes for primitives such as locks, barriers, and caches. For ZooKeeper operations like `create`, `delete`, `getData`, etc., it streamlines programming by allowing us to chain calls, a programming style often called *fluent*. It also provides namespaces, automatic reconnection, and other facilities that make applications more robust.

The Curator components were originally implemented and contributed by Netflix, and it has recently been promoted to a top-level project of the Apache Software Foundation.

In this chapter, we cover the implementation of the master in our example using Curator features. Our goal is not to provide a detailed and extensive discussion of Curator, but simply to introduce it and highlight some of the features that are convenient to use with a ZooKeeper application. Check the project page for an extensive list of its features.

## The Curator Client

Just as with ZooKeeper, before doing anything with Curator, we need to create a client. The client is typically an instance of `CuratorFramework` that we obtain by calling the Curator framework factory:

```
CuratorFramework zkc =
            CuratorFrameworkFactory.newClient(connectString, retryPolicy);
```

The `connectString` input parameter is the list of ZooKeeper servers we can connect to, just like when creating a ZooKeeper client. The `retryPolicy` parameter is a new feature of Curator. It enables the developer to specify a policy for retrying operations in the event of disconnections. Recall that with the regular ZooKeeper interface, we typically resubmit operations upon a connection loss event.

Our example instantiates the `CuratorFramework` client. There are other methods in the factory class to create an instance, but we don't cover them here. One is the `CuratorZooKeeperClient` class, which provides some additional functionality on top of the ZooKeeper client, such as enabling operations that are safe in the face of unanticipated disconnections. Unlike the `CuratorFramework` class, operations on a `CuratorZooKeeperClient` are executed directly against the ZooKeeper client handle.

## Fluent API

A fluent API enables us to write code by chaining calls instead of relying upon a rigid signature scheme for invoking an operation. For example, with the standard ZooKeeper API, we create a znode synchronously by invoking something like:

```
zk.create("/mypath",
          new byte[0],
          ZooDefs.Ids.OPEN_ACL_UNSAFE,
          CreateMode.PERSISTENT);
```

With the fluent API of Curator, we make the same call this way:

```
zkc.create().withMode(CreateMode.PERSISTENT).forPath("/mypath", new byte[0]);
```

The `create` call returns a `CreateBuilder` instance and the subsequent calls return an object of a type that `CreateBuilder` extends. For example, `CreateBuilder` extends `CreateModable<ACLBackgroundPathAndBytesable<String>>`, and `withMode` is declared in the generic `CreateModable<T>` interface. Builders are available for the other operations as well—`delete`, `setData`, `getData`, `checkExists`, and `getChildren`—through the Curator framework client object.

To execute the same operation asynchronously, we add `inBackground` as follows:

```
zkc.create().inBackground().withMode(CreateMode.PERSISTENT).forPath("/mypath",
    new byte[0]);
```

This returns immediately, and we have to create one or more listeners to receive the callback that is returned when the znode is created. We discuss listeners and how to register them in the next section.

There are a few different ways to implement the callback for an asynchronous call. If we issue the previous string of calls, the callback is delivered in the form of a `CREATE` event to registered listeners. The `inBackground` call optionally takes a context object, a concrete callback implementation to invoke, and even an executor (`java.util.concurrent.Executor`) to execute the callback. In Java, an executor is an object that executes runnable objects; we can use it here to decouple the execution of the callback from the callback thread of the ZooKeeper client. Using an executor is usually better than creating one new thread for each task.

To set a watch, we simply add `watched` to the call chain. For example:

```
zkc.getData().inBackground().watched().forPath("/mypath");
```

The notification triggered by the watcher is processed through listeners as well, and they are passed as a `WATCHED` event to a given listener. It is also possible to replace `watched` with a call to `usingWatcher`, which takes a regular ZooKeeper `Watcher` object and calls it when it receives the notification. A third option is to pass a `CuratorWatcher` object. The `process` method of `CuratorWatcher`, unlike from a ZooKeeper `Watcher`, might throw an exception.

## Listeners

Listeners process events that the Curator library generates. To exercise this mechanism, the application implements one or more listeners and registers them with the Curator framework client. Events are delivered to all registered listeners.

The listener mechanism is generic and can be used for all manner of events that happen asynchronously. As we discussed in the previous section, Curator uses listeners to process callbacks and watch notifications. The mechanism also can be used to handle the exceptions generated by background tasks.

Let's have a look at how to implement a listener that processes all callbacks and watch notifications for our master Curator example. The first step is to implement the template for a `CuratorListener`:

```
CuratorListener masterListener = new CuratorListener() {
    public void eventReceived(CuratorFramework client, CuratorEvent event) {
        try {
            switch (event.getType()) {
            case CHILDREN:
                ...

                break;
```

```
                case CREATE:
                    ...

                    break;
                case DELETE:
                    ...

                    break;
                case WATCHED:
                    ...

                    break;
                }
            } catch (Exception e) {
                LOG.error("Exception while processing event.", e);
                try {
                    close();
                } catch (IOException ioe) {
                    LOG.error("IOException while closing.", ioe);
                }
            }
        };
```

Because the goal here is just to illustrate the structure that we need to implement, we have omitted the code detail for each of the cases. Check the code examples that come with this book for more detail.

We next need to register the listener. For this we need a framework client, which we can create just like the first client we created:

```
client = CuratorFrameworkFactory.newClient(hostPort, retryPolicy);
```

Once we have the framework client, we register the listener as follows:

```
client.getCuratorListenable().addListener(masterListener);
```

A special kind of listener deals with errors reported when a background thread catches an exception. This is a low-level detail, but it might be necessary if you want to handle them in your application. When the application needs to deal with such errors, it must implement a different kind of listener:

```
UnhandledErrorListener errorsListener = new UnhandledErrorListener() {
    public void unhandledError(String message, Throwable e) {
        LOG.error("Unrecoverable error: " + message, e);
        try {
            close();
        } catch (IOException ioe) {
            LOG.warn( "Exception when closing.", ioe );
        }
    }
};
```

and register it with the listener client as follows:

```
client.getUnhandledErrorListenable().addListener(errorsListener);
```

Note that implementing listeners as event handlers, as we discussed in this section, is somewhat different from the way we proposed to implement ZooKeeper applications in previous chapters. For the master-worker example implemented directly on top of ZooKeeper (see "A Common Pattern" on page 72), we chain calls and callbacks, and each callback is handled by a different callback implementation. The callback implementations even have different types. With Curator, the details of a callback or a watch notification are encapsulated into an `Event` object, which makes it amenable to an implementation using a single event handler.

## State Changes in Curator

Curator exposes a different set of states than ZooKeeper. It has, for example, a `SUSPENDED` state, and it uses `LOST` to represent session expiration. The state machine for the connection states is illustrated in Figure 8-1. When dealing with state changes, our recommendation is in general to halt all operations of the master because we do not know if the ZooKeeper client will be able to reconnect before the session expires, and even if it does, the client might not be the primary master any more. It is safer to play conservatively in the case of a disconnection.

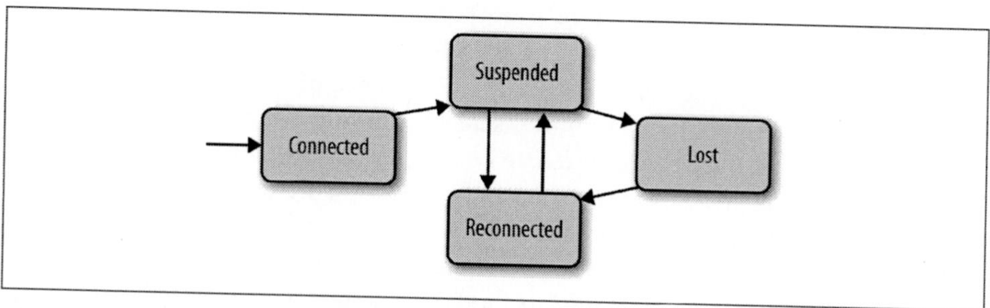


*Figure 8-1. Curator connection state machine*

There is an additional `READ_ONLY` state, which is not relevant for our example case. A connection goes into read-only mode if the ZooKeeper ensemble has read-only mode enabled and the server the client is connected to goes into read-only mode. As the server transitions to read-only mode, it cannot form a quorum with other servers because it is partitioned away. While the connection is in read-only mode, the client will miss any update that goes through. Such updates are possible if there is a subset of the ensemble that is able to form a quorum and that receives requests from the client to update the ZooKeeper state. A partition can last for arbitrarily long (it is out of the control of ZooKeeper) and consequently the number of updates it might miss is unbounded. Missing updates could lead to incorrect behavior of the application, so we strongly recommend thinking carefully about the consequences before enabling it. Note that the ability of going into read-only mode is not exclusive of Curator; ZooKeeper enables such an option (see Chapter 10).

## A Couple of Edge Cases

There are a couple of interesting error scenarios that Curator handles nicely. The first one has to do with the presence of errors during the creation of sequential znodes, and the second one with errors when deleting a znode:

*Sequential znodes*
: If the server the client is connected to crashes before returning the znode name (with the sequence number) or the client simply disconnects, then the client doesn't get a response even if the operation has been executed. As a consequence, the client doesn't know the path to the znode it created. Recall that we use sequential znodes, for example, in recipes that establish an order for participating clients. To address this problem, `CreateBuilder` provides a `withProtection` call that tells the Curator client to prefix the sequential znode with a unique identifier. If the `create` fails, the client retries the operation, and as part of retrying it verifies whether there is already a znode with the unique identifier.

*Guaranteed deletes*
: A similar situation occurs with `delete` operations. If the client disconnects from the server while executing a `delete` operation, it doesn't know whether the `delete` operation has succeeded or not. If the presence of the znode being deleted indicates, for example, that a resource is locked, it is important to delete the znode to make sure that the resource is free to be used again. The Curator client provides a call that enables an application to make the execution of a `delete` operation guaranteed. The operation is guaranteed in the sense that the Curator client reexecutes the operation until it succeeds, and for as long as the Curator client instance is valid. To use this feature, the `DeleteBuilder` interface defines a `guaranteed` call.

# Recipes

Curator provides a variety of recipes, and we encourage you to have a look at the extensive list of available recipes implemented. Here we discuss three recipes that we have used in the implementation of the Curator master: `LeaderLatch`, `LeaderSelector`, and `PathChildrenCache`.

## Leader Latch

We can use the leader latch primitive to elect a master in our application. First, we need a `LeaderLatch` instance:

```
leaderLatch = new LeaderLatch(client, "/master", myId);
```

The constructor of `LeaderLatch` takes a Curator framework client, a ZooKeeper path for this leadership group, and an identifier for this master. To enable callbacks when this Curator client acquires or loses leadership, we need to register an implementation of the `LeaderLatchListener` interface. This interface has two methods: `isLeader` and `notLeader`. This is what our `isLeader` implementation looks like:

```
@Override
public void isLeader()
{
...
    /*
     * Start workersCache ❶
     */
    workersCache.getListenable().addListener(workersCacheListener);
    workersCache.start();

    (new RecoveredAssignments(
      client.getZooKeeperClient().getZooKeeper())).recover(
        new RecoveryCallback() {
            public void recoveryComplete (int rc, List<String> tasks) {
                try {
                    if(rc == RecoveryCallback.FAILED) {
                        LOG.warn("Recovery of assigned tasks failed.");
                    } else {
                        LOG.info( "Assigning recovered tasks" );
                        recoveryLatch = new CountDownLatch(tasks.size());
                        assignTasks(tasks);    ❷
                    }


                    new Thread( new Runnable() { ❸
                        public void run() {
                            try {
                            /*
                             * Wait until recovery is complete
                             */
                            recoveryLatch.await();

                            /*
                             * Start tasks cache
                             */
                            tasksCache.getListenable().
                                addListener(tasksCacheListener); ❹
                            tasksCache.start();
                            } catch (Exception e) {
                                LOG.warn("Exception while assigning
                                         and getting tasks.",
                                         e  );
                            }
                        }
                   }).start();
```

```
                } catch (Exception e) {
                    LOG.error("Exception while executing the recovery callback",
                              e);
                }
            }
        });
    }
```

1. We start the workers cache before anything else to make sure that we have workers to assign tasks to.
2. Once we determine that we have tasks to assign that have not been assigned by the previous master, we proceed with assigning them.
3. We implement a barrier so that we wait until the assignment of recovered tasks ends before we move into assigning new tasks. If we don't do it, then the new master ends up assigning all recovered tasks again. Also, we do it in a separate thread just so that we don't lock the ZooKeeper client callback thread.
4. Once the master finishes with assigning recovered tasks, we start assigning new tasks.

We implement this method as part of the `CuratorMasterLatch` class, and `CuratorMasterLatch` implements `LeaderLatchListener`. We need to register the listener, however, before we actually start. We do both in the `runForMaster` method, on top of adding two other listeners for watch events and errors, respectively:

```
public void runForMaster() {
    client.getCuratorListenable().addListener(masterListener);
    client.getUnhandledErrorListenable().addListener(errorsListener);
    leaderLatch.addListener(this);
    leaderLatch.start();
}
```

For the `notLeader` call, which we execute once the master loses leadership, we simply close everything, which is sufficient for the purposes of this example. For a real application, you may need to clean up some local state and wait to become master again. If the `LeaderLatch` object is not closed, the Curator client will be considered for leadership again.

## Leader Selector

An alternative recipe for electing a master is `LeaderSelector`. The main difference between `LeaderLatch` and `LeaderSelector` is the listener interface they use. `LeaderSelector` uses `LeaderSelectorListener` instead, which defines a `takeLeadership` method and inherits `stateChanged`. We can use the leader latch primitive to elect a master in our application. First, we need a `LeaderSelector` instance:

```
leaderSelector = new LeaderSelector(client, "/master", this);
```

The constructor of `LeaderSelector` takes a Curator framework client, a ZooKeeper path for the leadership group this master is participating in, and an implementation of `LeaderSelectorListener`. The leadership group is the group of Curator clients participating in the master election. The `LeaderSelectorListener` implementation must contain both a `takeLeadership` method and a `stateChanged` one. The `takeLeadership` method is executed upon acquiring leadership, and most of its code for our example is the same as the code for `isLeader`. In our case, we implement it as follows:

```
CountDownLatch leaderLatch = new CountDownLatch(1);
CountDownLatch closeLatch = new CountDownLatch(1);

@Override
public void takeLeadership(CuratorFramework client) throws Exception
{
...
    /*
     * Start workersCache
     */
    workersCache.getListenable().addListener(workersCacheListener);
    workersCache.start();

    (new RecoveredAssignments(
      client.getZooKeeperClient().getZooKeeper())).recover(
        new RecoveryCallback() {
            public void recoveryComplete (int rc, List<String> tasks) {
                try {
                    if(rc == RecoveryCallback.FAILED) {
                        LOG.warn("Recovery of assigned tasks failed.");
                    } else {
                        LOG.info( "Assigning recovered tasks" );
                        recoveryLatch = new CountDownLatch(tasks.size());
                        assignTasks(tasks);
                    }


                    new Thread( new Runnable() {
                        public void run() {
                            try {
                            /*
                             * Wait until recovery is complete
                             */
                            recoveryLatch.await();

                            /*
                             * Start tasks cache
                             */
                            tasksCache.getListenable().
                                addListener(tasksCacheListener);
                            tasksCache.start();
                            } catch (Exception e) {
                                LOG.warn("Exception while assigning
```

```
                                          and getting tasks.",
                                          e );
                        }
                    }
                }).start();

                /*
                 * Decrement latch
                 */

                leaderLatch.countDown(); ❶
            } catch (Exception e) {
                LOG.error("Exception while executing the recovery callback",
                        e);
            }
        }
    });

    /*
     * This latch is to prevent this call from exiting. If we exit, then
     * we release mastership.
     */
    closeLatch.await(); ❷

}
```

❶ We provide a separate CountDownLatch to wait until this Curator client acquires leadership.

❷ If the master exits the takeLeadership call, it gives up mastership. We use a CountDownLatch to prevent it from exiting until we close the master.

We implement this method as part of the CuratorMaster class, and CuratorMaster implements LeaderSelectorListener. It is important that the master only exits takeLeadership if it wants to release mastership. We need, essentially, some form of lock to prevent it from exiting. In our implementation, we use a latch that we decrement when exiting the master instance.

We also start the leader selector in the runForMaster call, but unlike with LeaderLatch, we do not need to register a listener here (we register the listener in the constructor instead):

```
public void runForMaster() {
    client.getCuratorListenable().addListener(masterListener);
    client.getUnhandledErrorListenable().addListener(errorsListener);
    leaderSelector.setId(myId);
    leaderSelector.start();
}
```

We additionally give this master an arbitrary identifier. Although we have not done it in this example, we could also set the leader selector to automatically requeue (LeaderSelector.autoRequeue) upon losing leadership. Requeuing means that this client continuously tries to acquire leadership and it executes takeLeadership each time leadership is acquired.

As part of implementing the LeaderSelectorListener interface, we implement a method to handle connection state changes:

```
@Override
public void stateChanged(CuratorFramework client, ConnectionState newState)
{
    switch(newState) {
    case CONNECTED:
        //Nothing to do in this case.

        break;
    case RECONNECTED:
        // Reconnected, so I should ❶
        // still be the leader.

        break;
    case SUSPENDED:
        LOG.warn("Session suspended");

        break;
    case LOST:
        try {
            close(); ❷
        } catch (IOException e) {
            LOG.warn( "Exception while closing", e );
        }

        break;
    case READ_ONLY:
        // We ignore this case.

        break;
    }
}
```

❶ All operations of the master are through ZooKeeper. If the connection is lost, no operation of the master will go through. It is safe to do nothing.

❷ If the session is lost, we simply close this master.

## Children Cache

The last recipe we make use of in our example is the children cache (class PathChildrenCache). We use it both for the list of workers and for the list of tasks. This cache is responsible mainly for keeping a local copy of the list of children and for notifying us of changes to the cached set. Note that because of timing issues, the set might not be identical to the one ZooKeeper stores at a particular point in time, although it will eventually reflect changes to the ZooKeeper state.

To deal with changes for each instance of the cache, we implement the `PathChildrenCacheListener` interface, which has a single `childEvent` method. For the list of workers, we only care about workers going away because we need to reassign their tasks. Additions to the list are important when assigning new tasks:

```
PathChildrenCacheListener workersCacheListener = new PathChildrenCacheListener()
{
    public void childEvent(CuratorFramework client, PathChildrenCacheEvent event)
    {
        if(event.getType() == PathChildrenCacheEvent.Type.CHILD_REMOVED) {
            /*
             * Obtain just the worker's name
             */
            try {
                getAbsentWorkerTasks(event.getData().getPath().replaceFirst(
                                     "/workers/", ""));
            } catch (Exception e) {
                LOG.error("Exception while trying to re-assign tasks.", e);
            }
        }
    }
};
```

For the list of tasks, we use additions to the list to trigger the assignment process:

```
PathChildrenCacheListener tasksCacheListener = new PathChildrenCacheListener() {
    public void childEvent(CuratorFramework client, PathChildrenCacheEvent
                           event) {
        if(event.getType() == PathChildrenCacheEvent.Type.CHILD_ADDED) {
            try {
                assignTask(event.getData().getPath().replaceFirst("/tasks/",""));
            } catch (Exception e) {
                LOG.error("Exception when assigning task.", e);
            }
        }
    }
};
```

Note that we make an assumption here that there is at least one worker available to assign tasks to. In the case that there is no worker available, we need to hold the assignment by remembering the additions to the list that have not been assigned and assign them upon an addition to the list of workers. We do not implement this feature for the sake of simplicity; we leave it as an exercise for the reader.

## Takeaway Messages

Curator implements a set of nice extensions to the ZooKeeper API, abstracting away some of the complexities of ZooKeeper and implementing best practices gleaned from production experience and discussions in the community. In this chapter, we have covered how to leverage some of the features of Curator for the implementation of the master role in our master-worker example. We have particularly used the leader election implementations and the children cache to implement important features of the master. These two recipes are not the only ones Curator implements, however; a number of other recipes and features are available.

# PART III
# Administering ZooKeeper

This part of the book gives you the information you need to administer ZooKeeper. The internals provide the background that lets you make critical choices such as how many ZooKeeper servers to run and how to tune their communications.

CHAPTER 9

# ZooKeeper Internals

This chapter is a bit special compared to the others. It is not going to explicitly explain anything related to how to build applications with ZooKeeper. Instead, it explains how ZooKeeper works internally, by describing its protocols at a high level and the mechanisms it uses to tolerate faults while providing high performance. This content is important because it gives some deeper insight into why things work the way they work with ZooKeeper. This insight is important if you're planning on running ZooKeeper. It consequently serves as background for the next chapter.

As we saw in earlier chapters, ZooKeeper runs on an ensemble of servers while clients connect to these servers to execute operations. But what exactly are these servers doing with the operations the clients send? We hinted in Chapter 2 that we elect a distinguished server that we call the *leader*. The remaining servers, who follow the leader, are called *followers*. The leader is the central point for handling all requests that change the ZooKeeper system. It acts as a sequencer and establishes the order of updates to the ZooKeeper state. Followers receive and vote on the updates proposed by the leader to guarantee that updates to the state survive crashes.

The leader and the followers constitute the core entities guaranteeing the order of state updates despite crashes. There is a third kind of server, however, called an *observer*. Observers do not participate in the decision process of what requests get applied; they only learn what has been decided upon. Observers are there for scalability reasons.

In this chapter, we present the protocols we use to implement the ZooKeeper ensemble and the internals of servers and clients. We start with a discussion of some common concepts that we use throughout the remainder of the chapter regarding client requests and transactions.

**Code References**

Because this is a chapter about the internals, we figured that it might be interesting to provide references to the code, so that you can match the descriptions in this chapter to the source code. Pointers to classes and methods are provided where suitable.

## Requests, Transactions, and Identifiers

ZooKeeper servers process read requests (`exists`, `getData`, and `getChildren`) locally. When a server receives, say, a `getData` request from a client, it reads its state and returns it to the client. Because it serves requests locally, ZooKeeper is pretty fast at serving read-dominated workloads. We can add more servers to the ZooKeeper ensemble to serve more read requests, increasing overall throughput capacity.

Client requests that change the state of ZooKeeper (`create`, `delete`, and `setData`) are forwarded to the leader. The leader executes the request, producing a state update that we call a *transaction*. Whereas the request expresses the operation the way the client originates it, the transaction comprises the steps taken to modify the ZooKeeper state to reflect the execution of the request. Perhaps an intuitive way to explain this is to propose a simple, non-ZooKeeper operation. Say that the operation is `inc(i)`, which increments the value of the variable `i`. One possible request is consequently `inc(i)`. Say that the value of `i` is `10` and after incrementing it becomes `11`. Using the concepts of request and transaction, the request is `inc(i)` and the transaction is `i, 11` (variable `i` takes value `11`).

Let's now look at a ZooKeeper example. Say that a client submits a `setData` request on a given znode `/z`. `setData` should change the data of the znode and bump up the version number. So, a transaction for this request contains two important fields: the new data of the znode and the new version number of the znode. When applying the transaction, a server simply replaces the data of `/z` with the data in the transaction and the version number with the value in the transaction, rather than bumping it up.

A transaction is treated as a unit, in the sense that all changes it contains must be applied atomically. In the `setData` example, changing the data without an accompanying change to the version accordingly leads to trouble. Consequently, when a ZooKeeper ensemble applies transactions, it makes sure that all changes are applied atomically and there is no interference from other transactions. There is no rollback mechanism like with traditional relational databases. Instead, ZooKeeper makes sure that the steps of transactions do not interfere with each other. For a long time, the design used a single thread in each server to apply transactions. Having a single thread guarantees that the transactions are applied sequentially without interference. Recently, ZooKeeper has added support for multiple threads to speed up the process of applying transactions.

A transaction is also *idempotent*. That is, we can apply the same transaction twice and we will get the same result. We can even apply multiple transactions multiple times and get the same result, as long as we apply them in the same order every time. We take advantage of this idempotent property during recovery.

When the leader generates a new transaction, it assigns to the transaction an identifier that we call a *ZooKeeper transaction ID* (zxid). Zxids identify transactions so that they are applied to the state of servers in the order established by the leader. Servers also exchange zxids when electing a new leader, so they can determine which nonfaulty server has received more transactions and can synchronize their states.

A zxid is a `long` (64-bit) integer split into two parts: the *epoch* and the *counter*. Each part has 32 bits. The use of epochs and counters will become clear when we discuss Zab, the protocol we use to broadcast state updates to servers.

# Leader Elections

The leader is a server that has been chosen by an ensemble of servers and that continues to have support from that ensemble. The purpose of the leader is to order client requests that change the ZooKeeper state: `create`, `setData`, and `delete`. The leader transforms each request into a transaction, as explained in the previous section, and proposes to the followers that the ensemble accepts and applies them in the order issued by the leader.

To exercise leadership, a server must have support from a quorum of servers. As we discussed in Chapter 2, quorums must intersect to avoid the problem that we call *split brain*: two subsets of servers making progress independently. This situation leads to inconsistent system state, and clients end up getting different results depending on which server they happen to contact. We gave a concrete example of this situation in "ZooKeeper Quorums" on page 24.

The groups that elect and support a leader must intersect on at least one server process. We use the term *quorum* to denote such subsets of processes. Quorums pairwise intersect.

**Progress**

Because a quorum of servers is necessary for progress, ZooKeeper cannot make progress in the case that enough servers have permanently failed that no quorum can be formed. It is OK if servers are brought down and eventually boot up again, but for progress to be made, a quorum must eventually boot up. We relax this constraint when we discuss the possibility of reconfiguring ensembles in the next chapter. Reconfiguration can change quorums over time.

Each server starts in the LOOKING state, where it must either elect a new leader or find the existing one. If a leader already exists, other servers inform the new one which server is the leader. At this point, the new server connects to the leader and makes sure that its own state is consistent with the state of the leader.

If an ensemble of servers, however, are all in the LOOKING state, they must communicate to elect a leader. They exchange messages to converge on a common choice for the leader. The server that wins this election enters the LEADING state, while the other servers in the ensemble enter the FOLLOWING state.

The leader election messages are called *leader election notifications*, or simply *notifications*. The protocol is extremely simple. When a server enters the LOOKING state, it sends a batch of notification messages, one to each of the other servers in the ensemble. The message contains its current *vote*, which consists of the server's identifier (*sid*) and the zxid (*zxid*) of the most recent transaction it executed. Thus, (1,5) is a vote sent by the server with a sid of 1 and a most recent zxid of 5. (For the purposes of leader election, a zxid is a single number, but in some other protocols it is represented as an epoch and a counter.)

Upon receiving a vote, a server changes its vote according to the following rules:

1. Let voteId and voteZxid be the identifier and the zxid in the current vote of the receiver, whereas myZxid and mySid are the values of the receiver itself.
2. If (voteZxid > myZxid) or (voteZxid = myZxid and voteId > mySid), keep the current vote.
3. Otherwise, change my vote by assigning myZxid to voteZxid and mySid to vote Zxid.

In short, the server that is most up to date wins, because it has the most recent zxid. We'll see later that this simplifies the process of restarting a quorum when a leader dies. If multiple servers have the most recent zxid, the one with the highest sid wins.

Once a server receives the same vote from a quorum of servers, the server declares the leader elected. If the elected leader is the server itself, it starts executing the leader role. Otherwise, it becomes a follower and tries to connect to the elected leader. Note that it is not guaranteed that the follower will be able to connect to the elected leader. The elected leader might have crashed, for example. Once it connects, the follower and the leader sync their state, and only after syncing can the follower start processing new requests.

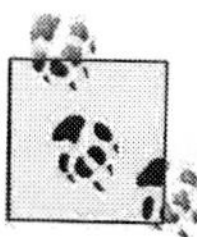

**Looking for a Leader**

The Java class in ZooKeeper that implements an election is `QuorumPeer`. Its `run` method implements the main loop of the server. When in the `LOOKING` state, it executes `lookForLeader` to elect a leader. This method basically executes the protocol we have just discussed. Before returning, the method sets the state of the server to either `LEADING` or `FOLLOWING`. `OBSERVING` is also an option that will be discussed later. If the server is leading, it creates a new `Leader` and runs it. If it is following, it creates a new `Follower` and runs it.

Let's go over an example of an execution of this protocol. Figure 9-1 shows three servers, each starting with a different initial vote corresponding to the server identifier and the last zxid of the server. Each server receives the votes of the other two, and after the first round, servers $s_2$ and $s_3$ change their votes to `(1,6)`. Servers $s_2$ and $s_3$ send a new batch of notifications after changing their votes, and after receiving these new notifications, each server has notifications from a quorum with the same vote. They consequently elect server $s_1$ to be the leader.

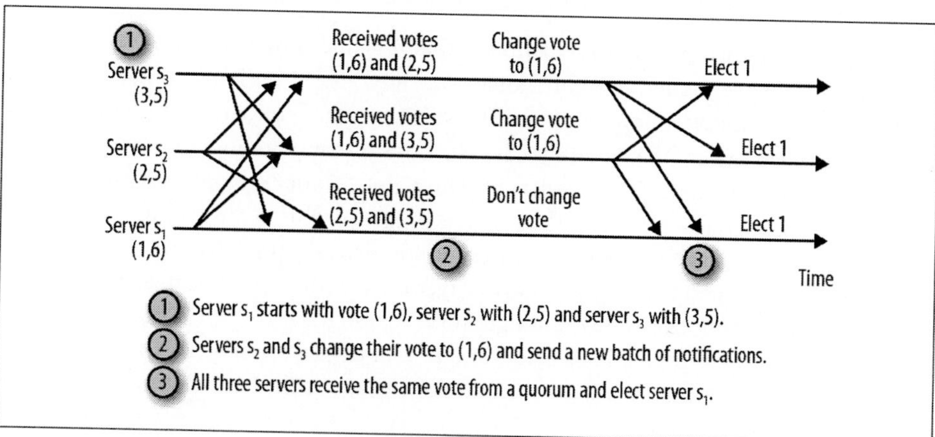


*Figure 9-1. Example of a leader election execution*

Not all executions are as well behaved as the one in Figure 9-1. In Figure 9-2, we show an example in which $s_2$ makes an early decision and elects a different leader from servers $s_1$ and $s_3$. This happens because the network happens to introduce a long delay in delivering the message from $s_1$ to $s_2$ that shows that $s_1$ has the higher zxid. In the meantime, $s_2$ elects $s_3$. In consequence, $s_1$ and $s_3$ will form a quorum, leaving out $s_2$.

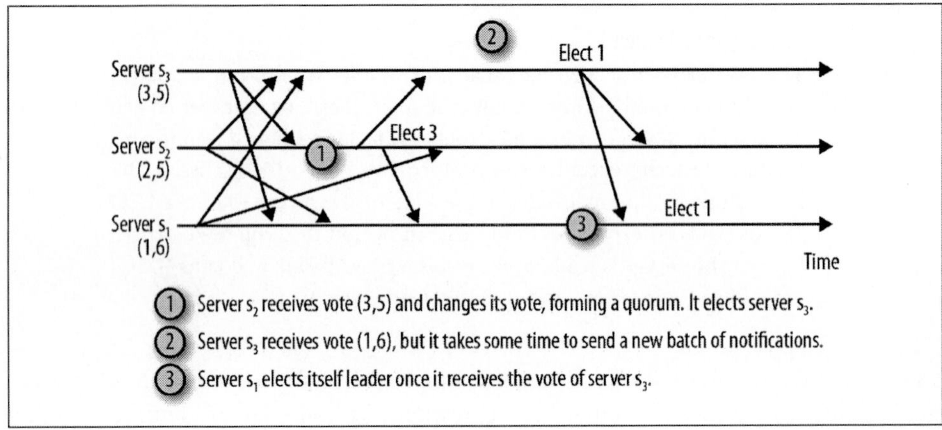


*Figure 9-2. Interleaving of messages causes a server to elect a different leader*

Having $s_2$ elect a different leader does not cause the service to behave incorrectly, because $s_3$ will not respond to $s_2$ as leader. Eventually $s_2$ will time out trying to get a response from its elected leader, $s_3$, and try again. Trying again, however, means that during this time $s_2$ will not be available to process client requests, which is undesirable.

One simple observation from this example is that if $s_2$ had waited a bit longer to elect a leader, it would have made the right choice. We show this situation in Figure 9-3. It is hard to know how much time a server should wait, though. The current implementation of `FastLeaderElection`, the default leader election implementation, uses a fixed value of 200 ms (see the constant `finalizeWait`). This value is longer than the expected message delay in modern data centers (less than a millisecond to a few milliseconds), but not long enough to make a substantial difference to recovery time. In case this delay (or any other chosen delay) is not sufficiently long, one or more servers will end up falsely electing a leader that does not have enough followers, so the servers will have to go back to leader election. Falsely electing a leader might make the overall recovery time longer because servers will connect and sync unnecessarily, and still need to send more messages to elect another leader.

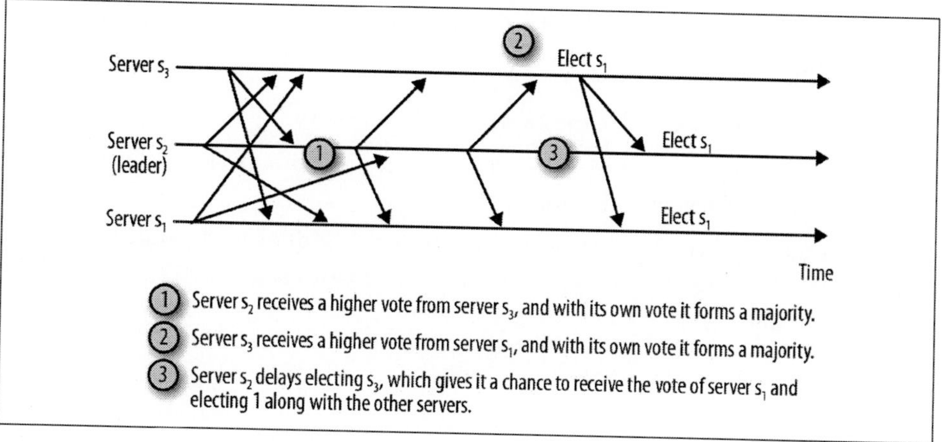


*Figure 9-3. Longer delay in electing a leader*

**What's Fast about Fast Leader Election?**

If you are wondering about it, we call the current default leader election algorithm *fast* for historical reasons. The initial leader election algorithm implemented a pull-based model, and the interval for a server to pull votes was about 1 second. This approach added some delay to recovery. With the current implementation, we are able to elect a leader faster.

To implement a new leader election algorithm, we need to implement the `Election` interface in the *quorum* package. To enable users to choose among the leader election implementations available, the code uses simple integer identifiers (see `QuorumPeer.createElectionAlgorithm()`). The other two implementations available currently are `LeaderElection` and `AuthFastLeaderElection`, but they have been deprecated as of release 3.4.0, so in some future releases you may not even find them.

# Zab: Broadcasting State Updates

Upon receiving a write request, a follower forwards it to the leader. The leader executes the request speculatively and broadcasts the result of the execution as a state update, in the form of a transaction. A transaction comprises the exact set of changes that a server must apply to the data tree when the transaction is committed. The data tree is the data structure holding the ZooKeeper state (see `DataTree`).

The next question to answer is how a server determines that a transaction has been committed. This follows a protocol called *Zab*: the ZooKeeper Atomic Broadcast protocol. Assuming that there is an active leader and it has a quorum of followers supporting its leadership, the protocol to commit a transaction is very simple, resembling a two-phase commit:

1. The leader sends a PROPOSAL message, $p$, to all followers.
2. Upon receiving $p$, a follower responds to the leader with an ACK, informing the leader that it has accepted the proposal.
3. Upon receiving acknowledgments from a quorum (the quorum includes the leader itself), the leader sends a message informing the followers to COMMIT it.

Figure 9-4 illustrates this sequence of steps. In the figure, we assume that the leader implicitly sends messages to itself.

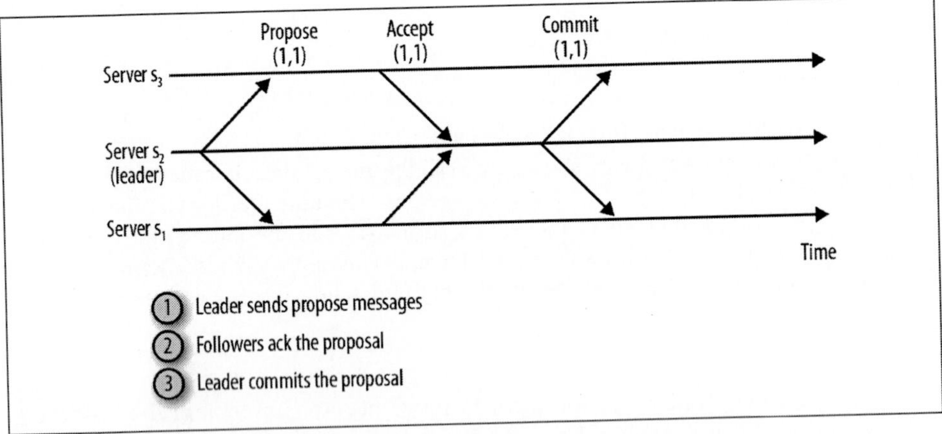


*Figure 9-4. Regular message pattern to commit proposals*

Before acknowledging a proposal, the follower needs to perform a couple of additional checks. The follower needs to check that the proposal is from the leader it is currently following, and that it is acknowledging proposals and committing transactions in the same order that the leader broadcasts them in.

Zab guarantees a couple of important properties:

- If the leader broadcasts $T$ and $T'$ in that order, each server must commit $T$ before committing $T'$.
- If any server commits transactions $T$ and $T'$ in that order, all other servers must also commit $T$ before $T'$.

The first property guarantees that transactions are delivered in the same order across servers, whereas the second property guarantees that servers do not skip transactions. Given that the transactions are state updates and each state update depends upon the previous state update, skipping transactions could create inconsistencies. The two-phase commit guarantees the ordering of transactions. Zab records a transaction in a quorum of servers. A quorum must acknowledge a transaction before the leader commits it, and a follower records on disk the fact that it has acknowledged the transaction.

As we'll see in "Local Storage" on page 170, transactions can still end up on some servers and not on others, because servers can fail while trying to write a transaction to storage. ZooKeeper can bring all servers up to date whenever a new quorum is created and a new leader chosen.

ZooKeeper, however, does not expect to have a single active leader the whole time. Leaders may crash or become temporarily disconnected, so servers may need to move to a new leader to guarantee that the system remains available. The notion of *epochs* represents the changes in leadership over time. An epoch refers to the period during which a given server exercised leadership. During an epoch, a leader broadcasts proposals and identifies each one according to a counter. Remember that each zxid includes the epoch as its first element, so each zxid can easily be associated to the epoch in which the transaction was created.

The epoch number increases each time a new leader election takes place. The same server can be the leader for different epochs, but for the purposes of the protocol, a server exercising leadership in different epochs is perceived as a different leader. If a server *s* has been the leader of epoch *4* and is currently the established leader of epoch *6*, a follower following *s* in epoch *6* processes only the messages *s* sent during epoch *6*. The follower may accept proposals from epoch *4* during the recovery period of epoch *6*, before it starts accepting new proposals for epoch *6*. Such proposals, however, are sent as part of epoch *6*'s messages.

Recording accepted proposals in a quorum is critical to ensure that all servers eventually commit transactions that have been committed by one or more servers, even if the leader crashes. Detecting perfectly that leaders (or any server) have crashed is very hard, if not impossible, in many settings, so it is very possible to falsely suspect that a leader has crashed.

Most of the difficulty with implementing a broadcast protocol is related to the presence of concurrent leaders, not necessarily in a split-brain scenario. Multiple concurrent leaders could make servers commit transactions out of order or skip transactions altogether, which leaves servers with inconsistent states. Preventing the system from ever having two servers believing they are leaders concurrently is very hard. Timing issues and dropped messages might lead to such scenarios, so the broadcast protocol cannot rely on this assumption. To get around this problem, Zab guarantees that:

- An elected leader has committed all transactions that will ever be committed from previous epochs before it starts broadcasting new transactions.
- At no point in time will two servers have a quorum of supporters.

To implement the first requirement, a leader does not become active until it makes sure that a quorum of servers agrees on the state it should start with for the new epoch. The initial state of an epoch must encompass all transactions that have been previously committed, and possibly some other ones that had been accepted before but not committed. It is important, though, that before the leader makes any new proposals for epoch *e*, it commits all proposals that will ever be committed from epochs up to and including $e - 1$. If there is a proposal lying around from epoch $e' < e$ and it is not committed by the leader of *e* by the time it makes the first proposal of *e*, the old proposal is never committed.

The second point is somewhat tricky because it doesn't really prevent two leaders from making progress independently. Say that a leader *l* is leading and broadcasting transactions. At some point, a quorum of servers *Q* believes *l* is gone, and it elects a new leader, *l′*. Let's say that *T* is a transaction that was being broadcast at the time *Q* abandoned *l*, and that a strict subset of *Q* has successfully recorded *T*. After *l′* is elected, enough processes not in *Q* also record *T*, forming a quorum for *T*. In this case, *T* is committed even after *l′* has been elected. But don't worry; this is not a bug. Zab guarantees that *T* is part of the transactions committed by *l′*, by guaranteeing that the quorum of supporters of *l′* contain at least one follower that has acknowledged *T*. The key point here is that *l′* and *l* do not have a quorum of supporters simultaneously.

Figure 9-5 illustrates this scenario. In the figure, *l* is server $s_5$, *l′* is $s_3$, *Q* comprises $s_1$ through $s_3$, and the zxid of *T* is ⟨1,1⟩. After receiving the second confirmation, $s_5$ is able to send a commit message to $s_4$ to tell it to commit the transaction. The other servers ignore messages from $s_5$ once they start following $s_3$. Note that $s_3$ acknowledged ⟨1,1⟩, so it is aware of the transaction when it establishes leadership.

We have just promised that Zab ensures the new leader *l′* does not miss ⟨1,1⟩, but how does it happen exactly? Before becoming active, the new leader *l′* must learn all proposals that servers in the old quorum have accepted previously, and it must get a promise that these servers won't accept further proposals from previous leaders. In the example in Figure 9-5, the servers forming a quorum and supporting *l′* promise that they won't accept any more proposals from leader *l*. At that point, if leader *l* is still able to commit any proposal, as it does with ⟨1,1⟩, the proposal must have been accepted by at least one server in the quorum that made the promise to the new leader. Recall that quorums must overlap in at least one server, so the quorum that *l* uses to commit and the quorum that *l′* talks to must have at least one server in common. Consequently, *l′* includes ⟨1,1⟩ in its state and propagates it to its followers.

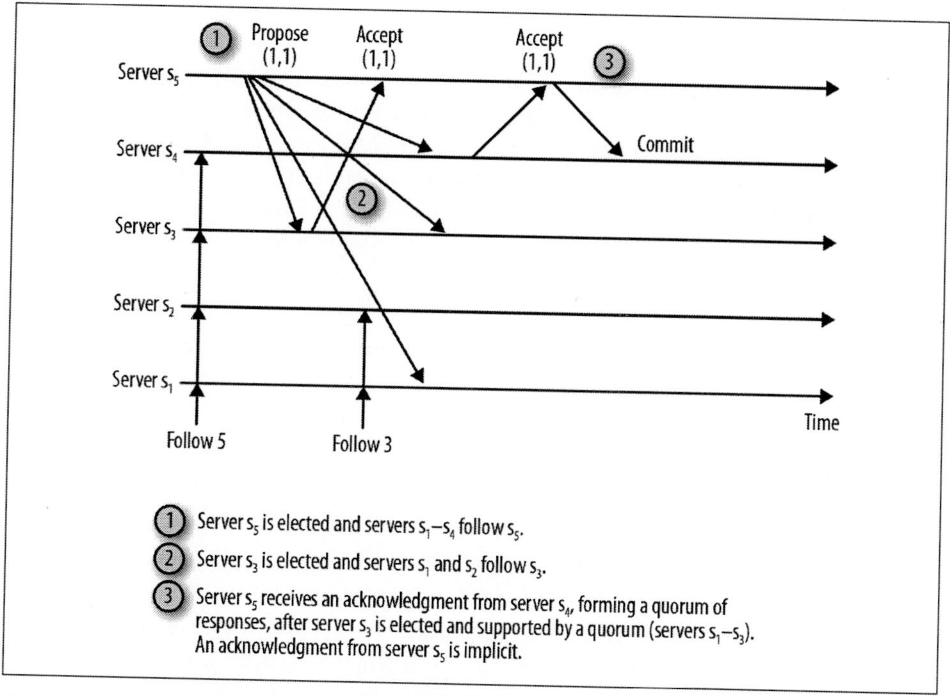


*Figure 9-5. Leaders overlapping illustration*

Recall that when electing a leader, servers pick the one with the highest zxid. This saves ZooKeeper from having to transfer proposals from followers to the leader; it only needs to transfer state from the leader to the followers. Say instead that we have at least one follower that has accepted a proposal that the leader hasn't. Before syncing up with the other followers, the leader would have to receive and accept the proposal. However, if we pick the server with the highest zxid, then we can completely skip this step and jump directly into bringing the followers up to date.

When transitioning between epochs, ZooKeeper uses two different ways to update the followers in order to optimize the process. If the follower is not too far behind the leader, the leader simply sends the missing transactions. They are always the most recent transactions, because followers accept all transactions in strict order. This update is called a `DIFF` in the code. If the follower is lagging far behind, ZooKeeper does a full snapshot transfer, called a `SNAP` in the code. Doing a full snapshot transfer increases recovery time, so sending a few missing transactions is preferable, but not always possible if the follower is far behind.

The `DIFF` a leader sends to a follower corresponds to the proposals that the leader has in its transaction log, whereas the `SNAP` is the latest valid snapshot that the leader has. Later in this chapter we discuss these two types of files that we keep on disk.

**Diving into the Code**

Here is a small guide to the code. Most of the Zab code is in `Leader`, `LearnerHandler`, and `Follower`. Instances of `Leader` and `LearnerHandler` are executed by the leader server, and `Follower` is executed by followers. Two important methods to look at are `Leader.lead` and `Follower.followLeader`. They are actually the methods executed when the servers transition from `LOOKING` to either `LEADING` or `FOLLOWING` in `QuorumPeer`.

For `DIFF` versus `SNAP`, follow the code in `LearnerHandler.run` to see how the code decides which proposals to send during a `DIFF`, and how snapshots are serialized and sent.

## Observers

We have focused so far on leaders and followers, but there is a third kind of server that we have not discussed: *observers*. Observers and followers have some aspects in common. In particular, they commit proposals from the leader. Unlike followers, though, observers do not participate in the voting process we discussed earlier. They simply *learn* the proposals that have been committed via `INFORM` messages. Both followers and observers are called *learners* because the leader tells them about changes of state.

**Rationale Behind INFORM Messages**

Because observers do not vote to accept a proposal, a leader does not send proposals to observers, and the commit messages that leaders send to followers do not contain the proposal itself, only its zxid. Consequently, just sending the commit message to an observer does not enable the observer to apply the proposal. That's the reason for using `INFORM` messages, which are essentially commit messages containing the proposals being committed.

In short, followers get two messages whereas observers get just one. Followers get the content of the proposal in a broadcast, followed by a simple commit message that has just the zxid. In contrast, observers get a single `INFORM` message with the content of the committed proposal.

Servers that participate in the vote that decides which proposals are committed are called `PARTICIPANT` servers. A `PARTICIPANT` server can be either a leader or a follower. Observers, in contrast, are called `OBSERVER` servers.

One main reason for having observers is scalability of read requests. By adding more observers, we can serve more read traffic without sacrificing the throughput of writes.

Note that the throughput of writes is driven by the quorum size. If we add more servers that can vote, we end up with larger quorums, which reduces write throughput. Adding observers, however, is not completely free of cost; each new observer induces the cost of one extra message per committed transaction. This cost is less, however, than that of adding servers to the voting process.

Another reason for observers is to have a deployment that spans multiple data centers. Scattering participants across data centers might slow down the system significantly because of the latency of links connecting data centers. With observers, update requests can be executed with high throughput and low latency in a single data center, while propagating to other data centers so that clients in other locations can consume them. Note that the use of observers does not eliminate network messages across data centers, because observers have to both forward update requests to the leader and process `INFORM` messages. It instead enables the messages necessary to commit updates to be exchanged in a single data center when all participants are set to run in the same data center.

# The Skeleton of a Server

Leaders, followers, and observers are all ultimately servers. The main abstraction we use in the implementation of a server is the *request processor*. A request processor is an abstraction of the various stages in a processing pipeline, and each server implements a sequence of such request processors. We can think of each processor as an element adding to the processing of a request. After being processed by all processors in the pipeline of a server, a given request can be declared to have been fully processed.

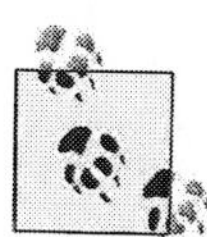

**Request Processors**

ZooKeeper code has an interface called `RequestProcessor`. The main method of the interface is `processRequest`, which takes a `Request` parameter. In a pipeline of request processors, the processing of requests for consecutive processors is usually decoupled using queues. When a processor has a request for the next processor, it queues the request, where it can wait until the next processor is ready to consume it.

## Standalone Servers

The simplest pipeline in ZooKeeper is for the standalone server (class `ZooKeeperServer`, no replication). Figure 9-6 shows the pipeline for this type of server. It has three request processors: `PrepRequestProcessor`, `SyncRequestProcessor`, and `FinalRequestProcessor`.

*Figure 9-6. Pipeline of a standalone server*

`PrepRequestProcessor` accepts a client request and executes it, generating a transaction as a result. Recall that the transaction is the result of executing an operation that is to be applied directly to the ZooKeeper data tree. The transaction data is added to the `Request` object in the form of a header and a transaction record. Also, note that only operations that change the state of ZooKeeper induce a transaction; read operations do not result in a transaction. The attributes referring to a transaction in a `Request` object are `null` for read requests.

The next request processor is `SyncRequestProcessor`. `SyncRequestProcessor` is responsible for persisting transactions to disk. It essentially appends transactions in order to a transaction log and generates snapshots frequently. We discuss disk state in more detail in the next section of this chapter.

The next and final processor is `FinalRequestProcessor`. It applies changes to the ZooKeeper data tree when the `Request` object contains a transaction. Otherwise, this processor reads the data tree and returns to the client.

## Leader Servers

When we switch to quorum mode, the server pipelines change a bit. Let's start with the leader pipeline (class `LeaderZooKeeperServer`), illustrated in Figure 9-7.

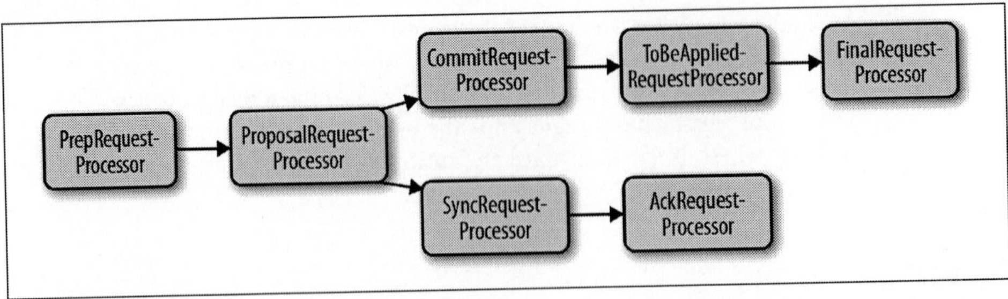


*Figure 9-7. Pipeline of a leader server*

The first processor is still `PrepRequestProcessor`, but the following processor now becomes `ProposalRequestProcessor`. It prepares proposals and sends them to the followers. `ProposalRequestProcessor` forwards all requests to `CommitRequestProcessor`, and additionally forwards the write requests to `SyncRequestProcessor`.

SyncRequestProcessor works the same as it does for the standalone server, and persists transactions to disk. It ends by triggering AckRequestProcessor, a simple request processor that generates an acknowledgment back to itself. As we mentioned earlier, the leader expects acknowledgments from every server in the quorum, including itself. AckRequestProcessor takes care of this.

The other processor following ProposalRequestProcessor is CommitRequestProcessor. CommitRequestProcessor commits proposals that have received enough acknowledgments. The acknowledgments are actually processed in the Leader class (the Leader.processAck() method), which adds committed requests to a queue in CommitRequestProcessor. The request processor thread processes this queue.

The next and final processor is FinalRequestProcessor, which is the same as the one used for the standalone server. FinalRequestProcessor applies update requests and executes read requests. Before FinalRequestProcessor, there stands a simple request processor that removes elements of a list of proposals to be applied. This request processor is called ToBeAppliedRequestProcessor. The list of to-be-applied requests contains requests that have been acknowledged by a quorum and are waiting to be applied. The leader uses this list to synchronize with followers and adds to this list when processing acknowledgments. ToBeAppliedRequestProcessor removes elements from this list after processing the request with FinalRequestProcessor.

Note that only update requests get into the to-be-applied list that ToBeAppliedRequestProcessor removes items from. ToBeAppliedRequestProcessor does not do any extra processing for read requests other than processing them with FinalRequestProcessor.

## Follower and Observer Servers

Let's talk now about followers (class FollowerRequestProcessor). Figure 9-8 shows the request processors a follower uses. Note that there isn't a single sequence of processors and that inputs come in different forms: client requests, proposals, and commits. We use arrows to specify the different paths a follower takes.

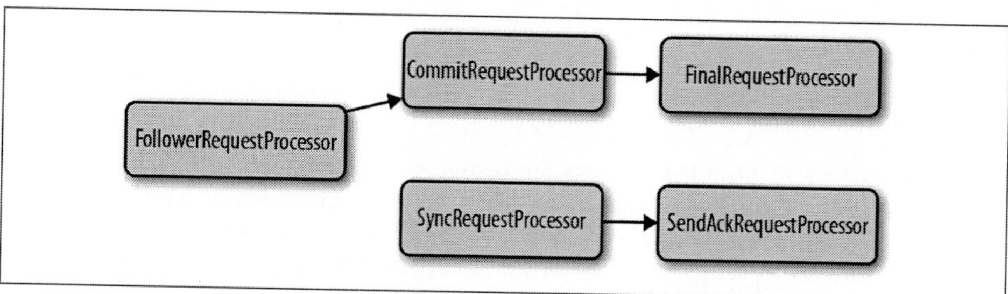


*Figure 9-8. Pipeline of a follower server*

We start with FollowerRequestProcessor, which receives and processes client requests. FollowerRequestProcessor forwards requests to CommitRequestProcessor, additionally forwarding write requests to the leader. CommitRequestProcessor forwards read requests directly to FinalRequestProcessor, whereas for write requests, CommitRequestProcessor must wait for a commit before forwarding to FinalRequest Processor.

When the leader receives a new write request, directly or through a learner, it generates a proposal and forwards it to followers. Upon receiving a proposal, a follower sends it to SyncRequestProcessor. SyncRequestProcessor processes the request, logging it to disk, and forwards it to SendAckRequestProcessor. SendAckRequestProcessor acknowledges the proposal to the leader. After the leader receives enough acknowledgments to commit a proposal, the leader sends commit messages to the followers (and sends INFORM messages to the observers). Upon receiving a commit message, a follower processes it with CommitRequestProcessor.

To guarantee that the order of execution is preserved, CommitRequestProcessor stalls the processing of pending requests once it encounters a write request. This means any read requests that were received after a write request will be blocked until the write request passes the CommitRequestProcessor. By waiting, it guarantees that requests are executed in the received order.

The request pipeline for observers (class ObserverZooKeeperServer) is very similar to the one for followers. But because observers do not need to acknowledge proposals, it is not necessary to send acknowledgment messages back to the leader or persist transactions to disk. Discussions are under way for making observers persist transactions to disk to speed up recovery for the observers, though. Consequently, future releases of ZooKeeper might have this feature.

## Local Storage

We have already mentioned transaction logs and snapshots, and that SyncRequestProcessor is the processor that writes them when processing write proposals. We'll focus a bit more on them in this section.

### Logs and Disk Use

Recall that servers use the transaction log to persist transactions. Before accepting a proposal, a server (follower or leader) persists the transaction in the proposal to the transaction log, a file on the local disk of the server to which transactions are appended in order. Every now and then, the server rolls over the log by closing the current file and creating a new one.

Because writing to the transaction log is in the critical path of write requests, ZooKeeper needs to be efficient about it. Appending to the file can be done efficiently on hard drives, but there are a couple of other tricks ZooKeeper uses to make it fast: group commits and padding. *Group commits* consist of appending multiple transactions in a single write to the disk. This allows many transactions to be persisted at the cost of a single disk seek.

There is one important caveat about persisting transactions to disk. Modern operating systems typically cache dirty pages and write them asynchronously to disk media. However, we need to make sure that transactions have been persisted before we move on. We consequently need to *flush* transactions onto disk media. Flushing here simply means that we tell the operating system to write dirty pages to disk and return when the operation completes. Because we persist transactions in `SyncRequestProcessor`, this processor is the one responsible for flushing. When it is time to flush a transaction to disk in `SyncRequestProcessor`, we in fact do it for all queued transactions to implement the group commit optimization. If there is one single transaction queued, the processor stills execute the flush. The processor does not wait for more queued transactions, which could increase the execution latency. For a code reference, check `SyncRequestProcessor.run()`.

**Disk Write Cache**

A server acknowledges a proposal only after forcing a write of the transaction to the transaction log. More precisely, the server calls the `commit` method of `ZKDatabase`, which ultimately calls `FileChannel.force`. This way, the server guarantees that the transaction has been persisted to disk before acknowledging it. There is a caveat to this observation, though. Modern disks have a write cache that stores data to be written to disk. If the write cache is enabled, a call to `force` does not guarantee that, upon return, the data is on media. Instead, it could be sitting in the write cache. To guarantee that written data is on media upon returning from a call to `FileChannel.force()`, the disk write cache must be disabled. Operating systems have different ways of disabling it.

*Padding* consists of preallocating disk blocks to a file. This is done so that updates to the file system metadata for block allocation do not significantly affect sequential writes to the file. If transactions are being appended to the log at a high speed, and if blocks were not preallocated to the file, the file system would need to allocate a new block whenever it reached the end of the one it was writing to. This would induce at least two extra disk seeks: one to update the metadata and another back to the file.

To avoid interference with other writes to the system, we strongly recommend that you write the transaction log to an independent device. A second device can be used for the operating system files and the snapshots.

## Snapshots

Snapshots are copies of the ZooKeeper data tree. Each server frequently takes a snapshot of the data tree by serializing the whole data tree and writing it to a file. The servers do not need to coordinate to take snapshots, nor do they have to stop processing requests. Because servers keep executing requests while taking a snapshot, the data tree changes as the snapshot is taken. We call such snapshots *fuzzy*, because they do not necessarily reflect the exact state of the data tree at any particular point in time.

Let's walk through an example to illustrate this. Say that a data tree has only two znodes: `/z` and `/z'`. Initially, the data of both `/z` and `/z'` is the integer 1. Now consider the following sequence of steps:

1. Start a snapshot.
2. Serialize and write `/z = 1` to the snapshot.
3. Set the data of `/z` to `2` (transaction *T*).
4. Set the data of `/z'` to `2` (transaction *T′*).
5. Serialize and write `/z' = 2` to the snapshot.

This snapshot contains `/z = 1` and `/z' = 2`. However, there has never been a point in time in which the values of both znodes were like that. This is not a problem, though, because the server replays transactions. It tags each snapshot with the last transaction that has been committed when the snapshot starts—call it *TS*. If the server eventually loads the snapshot, it replays all transactions in the transaction log that come after *TS*. In this case, they are *T* and *T′*. After replaying *T* and *T′* on top of the snapshot, the server obtains `/z = 2` and `/z' = 2`, which is a valid state.

An important follow-up question to ask is whether there is any problem with applying *T′* again because it had already been applied by the time the snapshot was taken. As we noted earlier, transactions are idempotent, so as long as we apply the same transactions in the same order, we will get the same result even if some of them have already been applied to the snapshot.

To understand this process, assume that applying a transaction consists of reexecuting the corresponding operation. In the case just described, the operation sets the data of the znode to a specific value, and the value is not dependent on anything else. Say that we are setting the data of `/z'` unconditionally (the version number is `-1` in the `setData` request). Reapplying the operation succeeds, but we end up with the wrong version number because we increment it twice. This can cause problems in the following way. Suppose that these three operations are submitted and executed successfully:

```
setData /z', 2, -1
setData /z', 3, 2
setData /a, 0, -1
```

The first `setData` operation is the same one we described earlier, but we've added two more `setData` operations to show that we can end up in a situation in which the second operation is not executed during a replay because of an incorrect version number. By assumption, all three requests were executed correctly when they were submitted. Suppose that a server loads the latest snapshot, which already contains the first `setData`. The server still replays the first `setData` operation because the snapshot is tagged with an earlier zxid. Because it reexecutes the first `setData`, the version does not match the one the second `setData` operation expects, so this operation does not go through. The third `setData` executes regularly because it is also unconditional.

After loading the snapshot and replaying the log, the state of the server is incorrect because it does not include the second `setData` request. This execution violates durability and the property that there are no gaps in the sequence of requests executed.

Such problems with reapplying requests are taken care of by turning transactions into *state deltas* generated by the leader. When the leader generates a transaction for a given request, as part of generating the transaction, it includes the changes in the request to the znode or its data and specifies a fixed version number. Reapplying a transaction consequently does not induce inconsistent version numbers.

# Servers and Sessions

Sessions constitute an important abstraction in ZooKeeper. Ordering guarantees, ephemeral znodes, and watches are tightly coupled to sessions. The session tracking mechanism is consequently very important to ZooKeeper.

One important task of ZooKeeper servers is to keep track of sessions. The single server tracks all sessions when running in standalone mode, whereas the leader tracks them in quorum mode. The leader server and the standalone server in fact run the same session tracker (see `SessionTracker` and `SessionTrackerImpl`). A follower server simply forwards session information for all the clients that connect to it to the leader (see `LearnerSessionTracker`).

To keep a session alive, a server needs to receive heartbeats for the session. Heartbeats come in the form of new requests or explicit ping messages (see `LearnerHandler.run()`). In both cases, the server *touches* sessions by updating the session expiration time (see `SessionTrackerImpl.touchSession()`). In quorum mode, a leader sends a `PING` message to learners and the learners send back the list of sessions that have been touched since the last `PING`. The leader sends a ping to learners every half a tick. A tick (described in "Basic Configuration" on page 179) is the minimum unit of time that ZooKeeper uses, expressed in milliseconds. So, if the tick is set to be 2 seconds, then the leader sends a ping every second.

Two important points govern session expiration. A data structure called the *expiry queue* (see `ExpiryQueue`) keeps session information for the purposes of expiration. The data structure keeps sessions in buckets, each bucket corresponding to a range of time during which the sessions are supposed to expire, and the leader expires the sessions in one bucket at a time. To determine which bucket to expire, if any, a thread checks the expiry queue to find out when the next deadline is. The thread sleeps until this deadline, and when it wakes up it polls the expiry queue for a new batch of sessions to expire. This batch can, of course, be empty.

To maintain the buckets, the leader splits time into `expirationInterval` units and assigns each session to the next bucket that expires after the session expiration time. The function doing the assignment essentially rounds the expiration time of a session up to the next higher interval. More concretely, the function evaluates this expression to determine which bucket a session belongs in when its session expiration time is updated:

```
(expirationTime / expirationInterval + 1) * expirationInterval
```

To provide an example, say that `expirationInterval` is 2 and the `expirationTime` for a given session occurs at time `10`. We assign this session to bucket 12 (the result of (10/2 + 1) * 2). Note that `expirationTime` keeps increasing as we touch the session, so we move the session to buckets that expire later accordingly.

One major reason for using a scheme of buckets is to reduce the overhead of checking for session expiration. A ZooKeeper deployment might have thousands of clients and consequently thousands of sessions. Checking for session expiration in a fine-grained manner is not suitable in such situations. Related to this comment, note that if the `expirationInterval` is short, ZooKeeper ends up performing session expiration checks in a fine-grained manner. The `expirationInterval` is currently one tick, which is typically on the order of seconds.

## Servers and Watches

Watches (see "Watches and Notifications" on page 20) are one-time triggers set by read operations, and each watch is triggered by a specific operation. To manage watches on the server side, a ZooKeeper server implements *watch managers*. An instance of the `WatchManager` class is responsible for keeping a list of current watches that are registered and for triggering them. All types of servers (standalone, leader, follower, and observer) process watches in the same way.

The `DataTree` class keeps a watch manager for child watches and another for data watches, the two types of watches discussed in "Getting More Concrete: How to Set Watches" on page 71. When processing a read operation that sets a watch, the class adds the watch to the manager's list of watches. Similarly, when processing a transaction, the class finds out whether any watches are to be triggered for the corresponding modification. If there are watches to be triggered, the class calls the trigger method of the manager. Both adding a watch and triggering a watch start with the execution of a read request or a transaction in `FinalRequestProcessor`.

A watch triggered on the server side is propagated to the client. The class responsible for this is the server `cnxn` object (see the `ServerCnxn` class), which represents the connection between the client and the server and implements the `Watcher` interface. The `Watcher.process` method serializes the watch event to a format that can be used to transfer it over the wire. The ZooKeeper client receives the serialized version of the watch event, transforms it back to a watch event, and propagates it to the application.

Watches are tracked only in memory. They are never persisted to the disk. When a client disconnects from a server, all its watches are removed from memory. Because client libraries also keep track of their outstanding watches, they will reestablish any outstanding watches on the new server that they connect with.

## Clients

There are two main classes in the client library: `ZooKeeper` and `ClientCnxn`. The `ZooKeeper` class implements most of the API, and this is the class a client application must instantiate to create a session. Upon creating a session, ZooKeeper associates a session identifier to it. The identifier is actually generated on the server side of the service (see `SessionTrackerImpl`).

The `ClientCnx` class manages the client socket connection with a server. It maintains a list of ZooKeeper servers it can connect to and transparently switches to a different server when a disconnection takes place. Upon reconnecting a session to a different server, the client also resets pending watches (see `ClientCnxn.SendThread.primeConnection()`). This reset is enabled by default, but can be disabled by setting `disableAutoWatchReset`.

## Serialization

For the serialization of messages and transactions to send over the network and to store on disk, ZooKeeper uses Jute, which grew out of Hadoop. Now the two code bases have evolved separately. Check the *org.apache.jute* package in the ZooKeeper code base for the Jute compiler code. (For a long time the ZooKeeper developer team has been discussing options for replacing Jute, but we haven't found a suitable replacement so far. It has served us well, though, and it hasn't been critical to replace it.)

The main definition file for Jute is *zookeeper.jute*. It contains all definitions of messages and file records. Here is an example of a Jute definition we have in this file:

```
module org.apache.zookeeper.txn {
...
```

```
    class CreateTxn {
        ustring path;
        buffer data;
        vector<org.apache.zookeeper.data.ACL> acl;
        boolean ephemeral;
        int parentCVersion;
    }
...
}
```

This example defines a module containing the definition of a `create` transaction. The module maps to a ZooKeeper package.

## Takeaway Messages

This chapter has discussed core ZooKeeper mechanisms. Leader election is critical for availability. Without it, a ZooKeeper ensemble cannot stay up reliably. Having a leader is necessary but not sufficient. ZooKeeper also needs the Zab protocol to propagate state updates, which guarantees a consistent state despite possible crashes of the ZooKeeper servers.

We have reviewed the types of servers: standalone, leader, follower, and observer. They differ in important ways with respect to the mechanisms they implement and the protocols they execute. Their use also has implications for a given deployment. For example, adding observers enables higher read throughput without affecting write throughput. Adding observers, however, does not increase the overall availability of the system.

Internally, ZooKeeper servers implement a number of mechanisms and data structures. Here we have focused on the implementation of sessions and watchers, important concepts to understand when implementing ZooKeeper applications.

Although we have provided pointers to the code in this chapter, the goal was not to provide an exhaustive view of the source code. We strongly encourage the reader to fetch a copy of the code and go over it, using the pointers here as starting points.

CHAPTER 10
# Running ZooKeeper

ZooKeeper was designed not only to be a great building block for developers, but also to be friendly for operations people. As distributed systems get bigger, managing operations becomes harder and robust administration practices become more important. Our vision was that ZooKeeper would be a standard distributed system component that an operations team could learn and manage well. We have seen from previous examples that a ZooKeeper server is easy to start up, but there are many knobs and dials to keep in mind when running a ZooKeeper service. Our goal in this chapter is to get you familiar and comfortable with the management tools available for running ZooKeeper.

In order for a ZooKeeper service to function correctly, it must be configured correctly. The distributed computing foundation upon which ZooKeeper is based works only when required operating conditions are met. For example, all ZooKeeper voting servers must have the same configuration. It has been our experience that improper or inconsistent configuration is the primary source of operational problems.

A simple example of one such problem happened in the early days of ZooKeeper. A team of early users had written their application around ZooKeeper, tested it thoroughly, and then pushed it to production. Even in the early days ZooKeeper was easy to work with and deploy, so this group pushed their ZooKeeper service and application into production without ever talking to us.

Shortly after the production traffic started, problems started appearing. We received a frantic call from operations saying that things were not working. He repeatedly asked them if they had thoroughly tested their solution before putting it into production, and they repeatedly assured him that they had. After piecing together the situation, he realized they were probably suffering from split brain. Finally he asked them to send him their configuration files. Once he saw them, it was clear what had happened: they had tested with a standalone ZooKeeper server, but when they went into production they used three servers to make sure they could tolerate a server failure. Unfortunately, they forgot to change the configuration, so they ended up pushing three standalone servers.

The clients treated all three servers as part of the same ensemble, but the servers themselves acted independently. Thus, three different groups of clients had three different (and conflicting) views of the system. It looked like everything was working fine, but behind the scenes it was chaos.

Hopefully, this example illustrates the importance of ZooKeeper configuration. It takes some understanding of basic concepts, but in reality it is not hard or complicated. The key is to know where the knobs are and what they do. That is what this chapter is about.

## Configuring a ZooKeeper Server

In this section we will look at the various knobs that govern how a ZooKeeper server operates. We have already seen a couple of these knobs, but there are many more. They all have default settings that correspond to the most common case, but often these should be changed. ZooKeeper was designed for easy use and operation, and we have succeeded so well that sometimes people get off and running without really understanding their setups. It is tempting to just go with the simplest configuration once everything starts working, but if you spend the time to learn the different configuration options, you will find that you can get better performance and more easily diagnose problems.

In this section we go through each of the configuration parameters, what they mean, and why you might need to use them. It may feel like a bit of a slog, so if you are looking for more exciting information, you may want to skip to the next section. If you do, though, come back at some point to sit down and familiarize yourself with the different options. It can make a big difference in the stability and performance of your ZooKeeper installation.

Each ZooKeeper server takes options from a configuration file named *zoo.cfg* when it starts. Servers that play similar roles and have the same basic setup can share a file. The *myid* file in the *data* directory distinguishes servers from each other. Each *data* directory must be unique to the server anyway, so it is a convenient place to put the distinguishing file. The server ID contained in the *myid* file serves as an index into the configuration file, so that a particular ZooKeeper server can know how it should configure itself. Of course, if servers have different setups (for example, they store their transaction logs in different places), each must have its own unique configuration file.

Configuration parameters are usually set in the configuration file. In the sections that follow, these parameters are presented in list form. Many parameters can also be set using Java system properties, which generally have the form `zookeeper.propertyName`. These properties are set using the *-D* option when starting the server. Where appropriate, the system property that corresponds to a given parameter will be presented in parentheses. A configuration parameter in a file has precedence over system properties.

## Basic Configuration

Some configuration parameters do not have a default and must be set for every deployment. These are:

`clientPort`
: The TCP port that clients use to connect to this server. By default, the server will listen on all of its interfaces for connections on this port unless `clientPortAddress` is set. The client port can be set to any number and different servers can listen on different ports. The default port is `2181`.

`dataDir` *and* `dataLogDir`
: `dataDir` is the directory where the fuzzy snapshots of the in-memory database will be stored. If this server is part of an ensemble, the *id* file will also be in this directory.

  The `dataDir` does not need to reside on a dedicated device. The snapshots are written using a background thread that does not lock the database, and the writes to storage are not synced until the snapshot is complete.

  Unless the `dataLogDir` option is set, the transaction log is also stored in this directory. The transaction log is very sensitive to other activity on the same device as this directory. The server tries to do sequential writes to the transaction log because the data must be synced to storage before the server can acknowledge a transaction. Other activity on the device—notably snapshots—can severely affect write throughput by causing disk heads to thrash during syncing. So, best practice is to use a dedicated log device and set `dataLogDir` to point to a directory on that device.

`tickTime`
: The length of a tick, measured in milliseconds. The tick is the basic unit of measurement for time used by ZooKeeper, and it determines the bucket size for session timeout as described in "Servers and Sessions" on page 173.

  The timeouts used by the ZooKeeper ensemble are specified in units of `tickTime`. This means, in effect, that the `tickTime` sets the lower bound on timeouts because the minimum timeout is a single tick. The minimum client session timeout is two ticks.

  The default `tickTime` is 3,000 milliseconds. Lowering the `tickTime` allows for quicker timeouts but also results in more overhead in terms of network traffic (heartbeats) and CPU time (session bucket processing).

## Storage Configuration

This section covers some of the more advanced configuration settings that apply to both standalone and ensemble configurations. They do not need to be set for ZooKeeper to function properly, but some (such as `dataLogDir`) really should be set.

`preAllocSize`
: The number of kilobytes to preallocate in the transaction log files (`zookeeper.preAllocSize`).

When writing to the transaction log, the server will allocate blocks of `preAllocSize` kilobytes at a time. This amortizes the file system overhead of allocating space on the disk and updating metadata. More importantly, it minimizes the number of seeks that need to be done.

By default, `preAllocSize` is 64 megabytes. One reason to lower this number is if the transaction log never grows that large. Because a new transaction log is restarted after each snapshot, if the number of transactions is small between each snapshot and the transactions themselves are small, 64 megabytes may be too big. For example, if we take a snapshot every 1,000 transactions, and the average transaction size is 100 bytes, a 100-kilobyte `preAllocSize` would be much more appropriate. The default `preAllocSize` is appropriate for the default `snapCount` and transactions that average more than 512 bytes in size.

`snapCount`
: The number of transactions between snapshots (`zookeeper.snapCount`).

When a ZooKeeper server restarts, it needs to restore its state. Two big factors in the time it takes to restore the state are the time it takes to read in a snapshot, and the time it takes to apply transactions that occurred after the snapshot was started. Snapshotting often will minimize the number of transactions that must be applied after the snapshot is read in. However, snapshotting does have an effect on the server's performance, even though snapshots are written in a background thread.

By default the `snapCount` is `100000`. Because snapshotting does affect performance, it would be nice if all of the servers in an ensemble were not snapshotting at the same time. As long as a quorum of servers is not snapshotting at once, the processing time should not be affected. For this reason, the actual number of transactions in each snapshot is a random number close to `snapCount`.

Note also that if `snapCount` is reached but a previous snapshot is still being taken, a new snapshot will not start and the server will wait another `snapCount` transactions before starting a new snapshot.

`autopurge.snapRetainCount`
: The number of snapshots and corresponding transaction logs to retain when purging data.

ZooKeeper snapshots and transaction logs are periodically garbage collected. The `autopurge.snapRetainCount` governs the number of snapshots to retain while garbage collecting. Obviously, not all of the snapshots can be deleted because that would make it impossible to recover a server; the minimum for `autopurge.snapRetainCount` is 3, which is also the default.

`autopurge.purgeInterval`
: The number of hours to wait between garbage collecting (purging) old snapshots and logs. If set to a nonzero number, `autopurge.purgeInterval` specifies the period of time between garbage collection cycles. If set to zero, the default, garbage collection will not be run automatically but should be run manually using the *zkCleanup.sh* script in the ZooKeeper distribution.

`fsync.warningthresholdms`
: The duration in milliseconds of a sync to storage that will trigger a warning (`fsync.warningthresholdms`).

  A ZooKeeper server will sync a change to storage before it acknowledges the change.

`weight.x=n`
: Used along with `group` options, this assigns a weight *n* to a server when forming quorums. The value *n* is the weight of a server when voting. A few parts of ZooKeeper require voting, such as leader election and the atomic broadcast protocol. By default, the weight of a server is 1. If the configuration defines groups but not weights, a weight of 1 will be assigned to all servers.

  If the sync system call takes too long, system performance can be severely impacted. The server tracks the duration of this call and will issue a warning if it is longer than `fsync.warningthresholdms`. By default, it's 1,000 milliseconds.

`traceFile`
: Keeps a trace of ZooKeeper operations by logging them in trace files named *traceFile.year.month.day*. Tracing is not done unless this option is set (`requestTraceFile`).

  This option is used to get a detailed view of the operations going through ZooKeeper. However, to do the logging, the ZooKeeper server must serialize the operations and write them to disk. This causes CPU and disk contention. If you use this option, be sure to avoid putting the trace file on the log device. Also realize that, unfortunately, tracing does perturb the system and thus may make it hard to recreate problems that happen when tracing is off. Just to make it interesting, the `traceFile` Java system property has no `zookeeper` prefix and the property name does not match the name of the configuration variable, so be careful.

## Network Configuration

These options place limits on communication between servers and clients. Timeouts are also covered in this section:

`globalOutstandingLimit`
: The maximum number of outstanding requests in ZooKeeper (`zookeeper.globalOutstandingLimit`).

ZooKeeper clients can submit requests faster than ZooKeeper servers can process them. This will lead to requests being queued at the ZooKeeper servers and eventually (as in, in a few seconds) cause the servers to run out of memory. To prevent this, ZooKeeper servers will start throttling client requests once the `globalOutstandingLimit` has been reached. But `globalOutstandingLimit` is not a hard limit; each client must be able to have at least one outstanding request, or connections will start timing out. So, after the `globalOutstandingLimit` is reached, the servers will read from client connections only if they do not have any pending requests.

To determine the limit of a particular server out of the global limit, we simply divide the value of this parameter by the number of servers. There is currently no smart way implemented to figure out the global number of outstanding operations and enforce the limit accordingly. Consequently, this limit is more of an upper bound on the number of outstanding requests. As a matter of fact, having the load perfectly balanced across servers is typically not achievable, so some servers that are running a bit slower or that are a bit more loaded may end up throttling even if the global limit has not been reached.

The default limit is 1,000 requests. You will probably not need to modify this parameter. If you have many clients that are sending very large requests you may need to lower the value, but we have never seen the need to change it in practice.

`maxClientCnxns`
: The maximum number of concurrent socket connections allowed from each IP address.

ZooKeeper uses flow control and limits to avoid overload conditions. The resources used in setting up a connection are much higher than the resources needed for normal operations. We have seen examples of errant clients that spun while creating many ZooKeeper connections per second, leading to a denial of service. To remedy the problem, we added this option, which will deny new connections from a given IP address if that address has `maxClientCnxns` active. The default is 60 concurrent connections.

Note that the connection count is maintained at each server. If we have an ensemble of five servers and the default is 60 concurrent connections, a rogue client will randomly connect to the five different servers and normally be able to establish close to 300 connections from a single IP address before triggering this limit on one of the servers.

`clientPortAddress`
: Limits client connections to those received on the given address.

  By default, a ZooKeeper server will listen on all its interfaces for client connections. However, some servers are set up with multiple network interfaces, generally one interface on an internal network and another on a public network. If you do not want a server to allow client connections from the public network, set the `clientPortAddress` to the address of the interface on the private network.

`minSessionTimeout`
: The minimum session timeout in milliseconds. When clients make a connection, they request a specific timeout, but the actual timeout they get will not be less than `minSessionTimeout`.

  ZooKeeper developers would love to be able to detect client failures immediately and accurately. Unfortunately, as "Building Distributed Systems with ZooKeeper" on page 7 explained, systems cannot do this under real conditions. Instead, they use heartbeats and timeouts. The timeouts to use depend on the responsiveness of the ZooKeeper client and server machines and, more importantly, the latency and reliability of the network between them. The timeout must be equal to at least the network round trip time between the client and server, but occasionally packets will be dropped, and when that happens the time it takes to receive a response is increased by the retransmission timeout as well as the latency of receiving the retransmitted packet.

  By default, `minSessionTimeout` is two times the `tickTime`. Setting this timeout too low will result in incorrect detection of client failures. Setting this timeout too high will delay the detection of client failures.

`maxSessionTimeout`
: The maximum session timeout in milliseconds. When clients make a connection, they request a specific timeout, but the actual timeout they get will not be greater than `maxSessionTimeout`.

  Although this setting does not affect the performance of the system, it does limit the amount of time for which a client can consume system resources. By default, `maxSessionTimeout` is 20 times the `tickTime`.

## Cluster Configuration

When an ensemble of servers provide the ZooKeeper service, we need to configure each server to have the correct timing and server list so that the servers can connect to each other and detect failures. These parameters must be the same on all the ZooKeeper servers in the ensemble:

`initLimit`
: The timeout, specified in number of ticks, for a follower to initially connect to a leader.

When a follower makes an initial connection to a leader, there can be quite a bit of data to transfer, especially if the follower has fallen far behind. `initLimit` should be set based on the transfer speed of the network between leader and follower and the amount of data to be transferred. If the amount of data stored by ZooKeeper is particularly large (i.e., if there are a large number of znodes or large data sets) or the network is particularly slow, `initLimit` should be increased. Because this value is so specific to the environment, there is no default for it. You should choose a value that will conservatively allow the largest expected snapshot to be transferred. Because you may have more than one transfer happening at a time, you may want to set `initLimit` to twice that expected time. If you set the `initLimit` too high, it will take longer for initial connections to faulty servers to fail, which can increase recovery time. For this reason it is a good idea to benchmark how long it takes for a follower to connect to a leader on your network with the amount of data you plan on using to find your expected time.

`syncLimit`
: The timeout, specified in number of ticks, for a follower to sync with a leader.

A follower will always be slightly behind the leader, but if the follower falls too far behind—due to server load or network problems, for example—it needs to be dropped. If the leader hasn't been able to sync with a follower for more than `syncLimit` ticks, it will drop the follower. Just like `initLimit`, `syncLimit` does not have a default and must be set. Unlike `initLimit`, `syncLimit` does not depend on the amount of data stored by ZooKeeper; instead, it depends on network latency and throughput. On high-latency networks it will take longer to send data and get responses back, so naturally the `syncLimit` will need to be increased. Even if the latency is relatively low, you may need to increase the `syncLimit` because any relatively large transaction may take a while to transmit to a follower.

`leaderServes`
: A "yes" or "no" flag indicating whether or not a leader will service clients (`zookeeper.leaderServes`).

The ZooKeeper server that is serving as leader has a lot of work to do. It talks with all the followers and executes all changes. This means the load on the leader is greater than that on the follower. If the leader becomes overloaded, the entire system may suffer.

This flag, if set to "no," can remove the burden of servicing client connections from the leader and allow it to dedicate all its resources to processing the change operations sent to it by followers. This will increase the throughput of operations that change system state. On the other hand, if the leader doesn't handle any of the client connections itself directly, the followers will have more clients because clients that would have connected to the leader will be spread among the followers. This is particularly problematic if the number of servers in an ensemble is low. By default, `leaderServes` is set to "yes."

`server.x=[hostname]:n:n[:observer]`
: Sets the configuration for server *x*.

  ZooKeeper servers need to know how to communicate with each other. A configuration entry of this form in the configuration file specifies the configuration for a given server *x*, where *x* is the ID of the server (an integer). When a server starts up, it gets its number from the *myid* file in the *data* directory. It then uses this number to find the `server.`*x* entry. It will configure itself using the data in this entry. If it needs to contact another server, *y*, it will use the information in the `server.`*y* entry to contact the server.

  The *`hostname`* is the name of the server on the network *n*. There are two TCP port numbers. The first port is used to send transactions, and the second is for leader election. The ports typically used are `2888:3888`. If `observer` is in the final field, the server entry represents an observer.

  Note that it is quite important that all servers use the same `server.`*x* configuration; otherwise, the ensemble won't work properly because servers might not be able to establish connections properly.

`cnxTimeout`
: The timeout value for opening a connection during leader election (`zookeeper.cnxTimeout`).

  The ZooKeeper servers connect to each other during leader election. This value determines how long a server will wait for a connection to complete before trying again. "Leader Elections" on page 157 showed the purpose of this timeout. The default value of 5 seconds is very generous and probably will not need to be adjusted.

`electionAlg`
: The election algorithm.

  We have included this configuration option for completeness. It selects among different leader election algorithms, but all have been deprecated except for the one that is the default. You shouldn't need to use this option.

## Authentication and Authorization Options

This section contains the options that are used for authentication and authorization. For infomation on configuration options for Kerberos, refer to "SASL and Kerberos" on page 113:

`zookeeper.DigestAuthenticationProvider.superDigest` *(Java system property only)*

This system property specifies the digest for the "super" user's password. (This feature is disabled by default.) A client that authenticates as `super` bypasses all ACL checking. The value of this system property will have the form `super:encoded_digest`. To generate the encoded digest, use the *org.apache.zookeeper.server.auth.DigestAuthenticationProvider* utility as follows:

```
java -cp $ZK_CLASSPATH \
    org.apache.zookeeper.server.auth.DigestAuthenticationProvider super:asdf
```

The following example command line generates an encoded digest for the password *asdf*:

```
super:asdf->super:T+4Qoey4ZZ8Fnni1Yl2GZtbH2W4=
```

To start a server using this digest, you can use the following command:

```
export SERVER_JVMFLAGS
SERVER_JVMFLAGS=-Dzookeeper.DigestAuthenticationProvider.superDigest=
                 super:T+4Qoey4ZZ8Fnni1Yl2GZtbH2W4=
./bin/zkServer.sh start
```

Now, when connecting with *zkCli*, you can issue the following:

```
[zk: localhost:2181(CONNECTED) 0] addauth digest super:asdf
[zk: localhost:2181(CONNECTED) 1]
```

At this point you are authenticated as the *super* user and will not be restricted by any ACLs.

**Unsecured Connections**

The connection between the ZooKeeper client and server is not encrypted, so the *super* password should not be used over untrusted links. The safest way to use the *super* password is to run the client using the *super* password on the same machine as the ZooKeeper server.

## Unsafe Options

The following options can be useful, but be careful when you use them. They really are for very special situations. The majority of administrators who think they need them probably don't:

`forceSync`
: A "yes" or "no" option that controls whether data should be synced to storage (`zookeeper.forceSync`).

By default, and when `forceSync` is set to `yes`, transactions will not be acknowledged until they have been synced to storage. The sync system call is expensive and is the cause of one of the biggest delays in transaction processing. If `forceSync` is set to `no`, transactions will be acknowledged as soon as they have been written to the operating system, which usually caches them in memory before writing them to disk. Setting `forceSync` to `no` will yield an increase in performance at the cost of recoverability in the case of a server crash or power outage.

`jute.maxbuffer` *(Java system property only)*
: The maximum size, in bytes, of a request or response. This option can be set only as a Java system property. There is no `zookeeper.` prefix on it.

ZooKeeper has some built-in sanity checks, one of which is the amount of data that can be transferred for a given znode. ZooKeeper is designed to store configuration data, which generally consists of small amounts of metadata information (on the order of hundreds of bytes). By default, if a request or response has more than 1 megabyte of data, it is rejected as insane. You may want to use this property to make the sanity check smaller or, if you really are insane, increase it.

**Changing the Sanity Check**

Although the size limit specified by `jute.maxbuffer` is most obviously exceeded with a large write, the problem can also happen when getting the list of children of a znode with a lot of children. If a znode has hundreds of thousands of immediate child znodes whose names average 10 characters in length, the default maximum buffer size will get hit when trying to return the list of children, causing the connection to get reset.

`skipACL`
: Skips all ACL checks (`zookeeper.skipACL`).

There is some overhead associated with ACL checking. This option can be used to turn off all ACL checking. It will increase performance, but will leave the data completely open to any client that can connect to a ZooKeeper server.

`readonlymode.enabled` *(Java system property only)*
: Setting this value to `true` enables read-only-mode server support. Clients that request read-only-mode support will be able to connect to a server to read (possibly stale) information even if that server is partitioned from the quorum. To enable read-only mode, a client needs to set `canBeReadOnly` to `true`.

This feature enables a client to read (but not write) the state of ZooKeeper in the presence of a network partition. In such cases, clients that have been partitioned away can still make progress and don't need to wait until the partition heals. It is very important to note that a ZooKeeper server that is disconnected from the rest of the ensemble might end up serving stale state in read-only mode.

## Logging

The ZooKeeper server uses SLF4J (the Simple Logging Facade for Java) as an abstraction layer for logging, and by default uses Log4J to do the actual logging. It may seem like overkill to use two layers of logging abstractions, and it is. In this section we will give a brief overview of how to configure Log4J. Although Log4J is very flexible and powerful, it is also a bit complicated. There are whole books written about it; in this section we will just give a brief overview of the basics needed to get things going.

The Log4J configuration file is named *log4j.properties* and is pulled from the classpath. One disappointing thing about Log4J is that if you don't have the *log4j.properties* file in place, you will get the following output:

```
log4j:WARN No appenders could be found for logger (org.apache.zookeeper.serv ...
log4j:WARN Please initialize the log4j system properly.
```

That's it; all the rest of the log messages will be discarded.

Generally, *log4j.properties* is kept in a *conf* directory that is included in the classpath. Let's look at the main part of the *log4j.properties* file that is distributed with ZooKeeper:

```
zookeeper.root.logger=INFO, CONSOLE ❶
zookeeper.console.threshold=INFO
zookeeper.log.dir=.
zookeeper.log.file=zookeeper.log
zookeeper.log.threshold=DEBUG
zookeeper.tracelog.dir=.
zookeeper.tracelog.file=zookeeper_trace.log

log4j.rootLogger=${zookeeper.root.logger} ❷

log4j.appender.CONSOLE=org.apache.log4j.ConsoleAppender ❸
log4j.appender.CONSOLE.Threshold=${zookeeper.console.threshold} ❹
log4j.appender.CONSOLE.layout=org.apache.log4j.PatternLayout ❺
log4j.appender.CONSOLE.layout.ConversionPattern=%d{ISO8601} [myid:%X{myid}] -
...

log4j.appender.ROLLINGFILE=org.apache.log4j.RollingFileAppender ❻
log4j.appender.ROLLINGFILE.Threshold=${zookeeper.log.threshold} ❼
log4j.appender.ROLLINGFILE.File=${zookeeper.log.dir}/${zookeeper.log.file}
log4j.appender.ROLLINGFILE.MaxFileSize=10MB
log4j.appender.ROLLINGFILE.MaxBackupIndex=10
log4j.appender.ROLLINGFILE.layout=org.apache.log4j.PatternLayout
```

```
log4j.appender.ROLLINGFILE.layout.ConversionPattern=%d{ISO8601} [myid:%X{myid}] -
...
```

❶ This first group of settings, which all start with `zookeeper.`, set up the defaults for this file. They are actually system properties and can be overridden with the corresponding *-D* JVM options on the Java command line. The first line configures logging. The default set here says that messages below the `INFO` level should be discarded and messages should be output using the `CONSOLE` appender. You can specify multiple appenders; for example, you could set `zookeeper.root.logger` to `INFO, CONSOLE, ROLLINGFILE` if you wanted to send log messages to both the `CONSOLE` and `ROLLINGFILE` appenders.

❷ The `rootLogger` is the logger that processes all log messages, because we do not define any other loggers.

❸ This line associates the name `CONSOLE` with the class that will actually be handling the output of the message—in this case, `ConsoleAppender`.

❹ Appenders can also filter messages. This line states that this appender will ignore any messages below the `INFO` level, because that is the threshold set in `zookeeper.root.logger`.

❺ Appenders use a layout class to format the messages before they are written out. We use the pattern layout to log the message level, date, thread information, and calling location information in addition to the message itself.

❻ The `RollingFileAppender` will implement rolling log files rather than continually appending to a single log or console. Unless `ROLLINGFILE` is referenced by the `rootLogger`, this appender will be ignored.

❼ The threshold for the `ROLLINGFILE` is set to `DEBUG`. Because the `rootLogger` filters out all messages below the `INFO` level, no `DEBUG` messages will get to the `ROLLINGFILE`. If you want to see the `DEBUG` messages, you must also change `INFO` to `DEBUG` in `zookeeper.root.logger`.

Logging can affect the performance of a process, especially at the `DEBUG` level. At the same time, logging can provide valuable information for diagnosing problems. A useful way to balance the performance cost of detailed logging with the insight that logging gives you is to set the appenders to have thresholds at `DEBUG` and set the level of the `rootLogger` to `WARN`. When the server is running, if you need to diagnose a problem as it is happening, you can change the level of the `rootLogger` to `INFO` or `DEBUG` on the fly with JMX to examine system activity more closely.

## Dedicating Resources

As you think about how you will configure ZooKeeper to run on a machine, it is also important to think about the machine itself. For good, consistent performance you will want to have a dedicated log device. This means the log directory must have its own hard drive that is not used by other processes. You don't even want ZooKeeper to use it for the periodic fuzzy snapshots it does.

You should also consider dedicating the whole machine on which it will run to ZooKeeper. ZooKeeper is a critical component that needs to be reliable. We can use replication to handle failures of ZooKeeper servers, so it is tempting to think that the machines that ZooKeeper runs on do not need to be particularly reliable and can be shared with other processes. The problem is that other processes can greatly increase the probability of a ZooKeeper failure. If another process starts making the disk thrash or uses all the memory or CPU, it will cause the ZooKeeper server to fail, or at least perform very poorly. A particularly problematic scenario is when a ZooKeeper server process runs on the same server as one of the application processes that it manages. If that process goes into an infinite loop or starts behaving badly, it may adversely affect the ZooKeeper server process at the very moment that it is needed to allow other processes to take over from the bad one. Dedicate a machine to run each ZooKeeper server in order to avoid these problems.

# Configuring a ZooKeeper Ensemble

The notion of quorums, introduced in "ZooKeeper Quorums" on page 24, is deeply embedded in the design of ZooKeeper. The quorum is relevant when processing requests and when electing a leader in replicated mode. If a quorum of ZooKeeper servers is up, the ensemble makes progress.

A related concept is that of observers, explained in "Observers" on page 166. Observers participate in the ensemble, receiving requests from clients and process state updates from the servers. The leader, however, does not consider observer acknowledgments when processing requests. The ensemble also does not consider observer notifications when electing a leader. Here we discuss how to configure quorums and ensembles.

## The Majority Rules

When an ensemble has enough ZooKeeper servers to start processing requests, we call the set of servers a *quorum*. Of course, we never want there to be two disjoint sets of servers that can process requests, or we would end up with split brain. We can avoid the split-brain problem by requiring that all quorums have at least a majority of servers. (Note: half of the servers do not constitute a majority; you must have greater than half the number of servers to have a majority.)

When we set up a ZooKeeper ensemble to use multiple servers, we use majority quorums by default. ZooKeeper automatically detects that it should run in replicated mode because there are multiple servers in the configuration (see "Cluster Configuration" on page 183), and it defaults to using majority quorums.

## Configurable Quorums

One important property we have mentioned about quorums is that, if one quorum is dissolved and another formed within the ensemble, the two quorums must intersect in at least one server. Majority quorums clearly satisfy this intersection property. Quorums in general, however, are not constrained to majorities, and ZooKeeper allows flexible configuration of quorums. The particular scheme we use is to group servers into disjoint sets and assign weights to the servers. To form a quorum in this scheme, we need a majority of votes from each of a majority of groups. For example, say that we have three groups, each group containing three servers, and each server having weight 1. To form a quorum in this example, we need four servers: two servers from one group and two servers from a different group. In general, the math boils down to the following. If we have $G$ groups, then we need servers from a subset $G'$ of servers such that $|G'| > |G|/2$. Additionally, for each $g$ in $G'$, we need a subset $g'$ of $g$ such that the sum of the weights $W'$ of $g'$ is at least half of the sum of the weights of $g$ (i.e., $W' > W/2$).

The following configuration option creates a group:

`group.x=n[:n]`
: Enables a hierarchical quorum construction. *x* is a group identifier and the numbers following the equals sign correspond to server identifiers. The right side of the assignment is a colon-separated list of server identifiers. Note that groups must be disjoint and the union of all groups must be the ZooKeeper ensemble. In other words, every server in the ensemble must be listed once in some group.

Here is an example of nine servers split across three different groups:

```
group.1=1:2:3
group.2=4:5:6
group.3=7:8:9
```

In this example all servers have the same weight, and to form a quorum we need two servers from two groups, or a total of four servers. With majority quorums, we would need at least five servers to form a quorum. Note that the quorum cannot be formed from any subset of four servers, however: an entire group plus a single server from a different group does not form a quorum.

A configuration like this has a variety of benefits when we are deploying ZooKeeper across different data centers. For example, one group may represent a group of servers running in a different data center, and if that data center goes down, ZooKeeper can keep going.

One way of deploying across three data centers that tolerates one data center going down and uses majorities is to have three servers in each of two data centers and put only one server in the third data center. If any of the data centers becomes unavailable, the other two can form a quorum. This configuration has the advantage that any four servers out of the seven form a quorum. One shortcoming is that the number of servers is not balanced across the data centers. A second shortcoming is that once a data center becomes unavailable, no further server crashes in other data centers are tolerated.

If there are only two data centers available, we can use weights to express a preference, based, for instance, on the number of ZooKeeper clients in each data center. With only two data centers, we can't tolerate either of the data centers going down if the servers all have equal weight, but we can tolerate one of the two going down if we assign a higher weight to one of the servers. Say that we assign three servers to each of the data centers and we include them all in the same group:

```
group.1=1:2:3:4:5:6
```

Because all the servers default to the same weight, we will have a quorum of servers as long as four of the six servers are up. Of course, this means that if one of the data centers goes down, we will not be able to form a quorum even if the three servers in the other data center are up.

To assign different weights to servers, we use the following configuration option:

`weight.x=n`
: Used along with `group` options, this assigns a weight *n* to a server when forming quorums. The value *n* is the weight the server has when voting. A few parts of ZooKeeper require voting, such as leader election and the atomic broadcast protocol. By default, the weight of a server is 1. If the configuration defines groups but not weights, a weight of 1 will be assigned to all servers.

Let's say that we want one of the data centers, which we will call *D1*, to still be able to function as long as all of its servers are up even if the other data center is down. We can do this by assigning one of the servers in *D1* more weight, so that it can more easily form a quorum with other servers.

Let's assume that servers 1, 2, and 3 are in *D1*. We use the following line to assign server 1 more weight:

```
weight.1=2
```

With this configuration, we have seven votes total and we need four votes to form a quorum. Without the `weight.1=2` parameter, any server needs at least three other servers to form a quorum, but with that parameter server 1 can form a quorum with just two other servers. So if *D1* is up, even if the other data center fails, servers 1, 2, and 3 can form a quorum and continue operation.

These are just a couple of examples of how different quorum configurations might impact a deployment. The hierarchical scheme we provide is flexible, and it enables other configurations with different weights and group organizations.

## Observers

Recall that observers are ZooKeeper servers that do not participate in the voting protocol that guarantees the order of state updates. To set up a ZooKeeper ensemble that uses observers, add the following line to the configuration files of any servers that are to be observers:

```
peerType=observer
```

You also need to add `:observer` to the server definition in the configuration file of each server, like this:

```
server.1:localhost:2181:3181:observer
```

# Reconfiguration

Wow, configuration is a lot of work, isn't it? But you've worked it all out, and now you have your ZooKeeper ensemble made up of three different machines. But then a month or two goes by, and you realize that the number of client processes using ZooKeeper has grown and it has become a much more mission-critical service. So you want to grow it to an ensemble of five machines. No big deal, right? You can pull the cluster down late one night, reconfigure everything, and have it all back up in less than a minute. Your users may not even see the outage if your applications handle the `Disconnected` event correctly. That's what we thought we when first developed ZooKeeper, but it turns out that things are more complicated.

Look at the scenario in Figure 10-1. Three machines (A, B, and C) make up the ensemble. C is lagging behind a bit due to some network congestion, so it has only seen transactions up to ⟨1,3⟩ (where 1 is the epoch and 3 is the transaction within that epoch, as described in "Requests, Transactions, and Identifiers" on page 156. But A and B are actively communicating, so C's lag hasn't slowed down the system. A and B have been able to commit up to transaction ⟨1,6⟩.

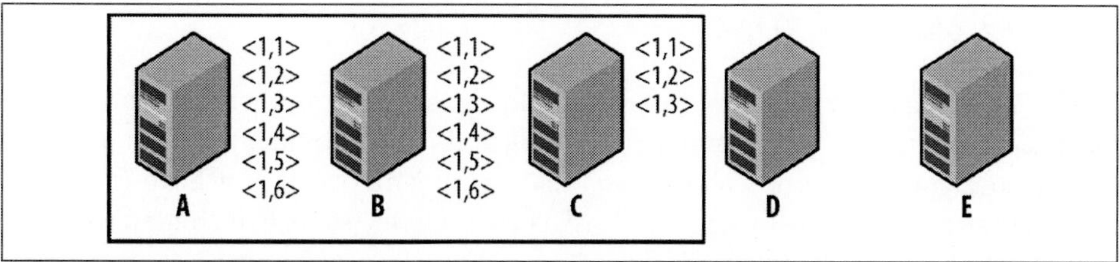


*Figure 10-1. An ensemble of three servers about to change to five*

Now suppose we bring down these machines to add D and E into the mix. Of course, the two new machines don't have any state at all. We reconfigure A, B, C, D, and E to be one big ensemble and start everything back up. Because we have five machines, we need three machines to form a quorum. C, D, and E are enough for a quorum, so in Figure 10-2 we see what happens when they form a quorum and sync up. This scenario can easily happen if A and B are slow at starting, perhaps because they were started a little after the other three. Once our new quorum syncs up, A and B will sync with C because it has the most up-to-date state. However, our three quorum members will each end up with ⟨1,3⟩ as the last transaction. They never see ⟨1,4⟩, ⟨1,5⟩, and ⟨1,6⟩, because the only two servers that did see them are not part of this new quorum.

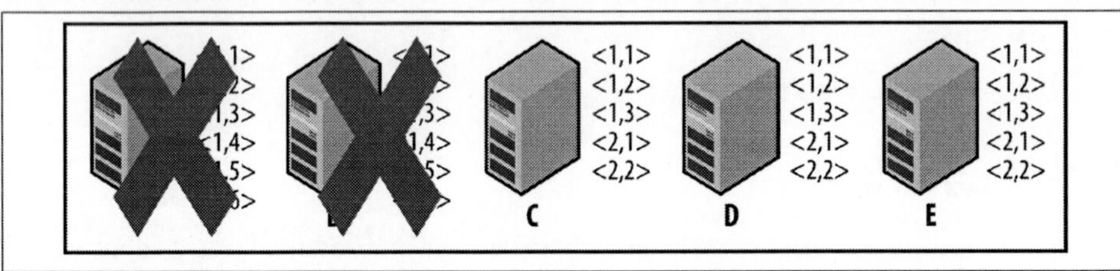


*Figure 10-2. An ensemble of five servers with a quorum of three*

Because we have an active quorum, these servers can actually commit new transactions. Let's say that two new transactions come in: ⟨2,1⟩ and ⟨2,2⟩. Now, in Figure 10-3, when A and B finally come up and connect with C, who is the leader of the quorum, C welcomes them in and promptly tells them to delete transactions ⟨1,4⟩, ⟨1,5⟩, and ⟨1,6⟩ while receiving ⟨2,1⟩ and ⟨2,2⟩.

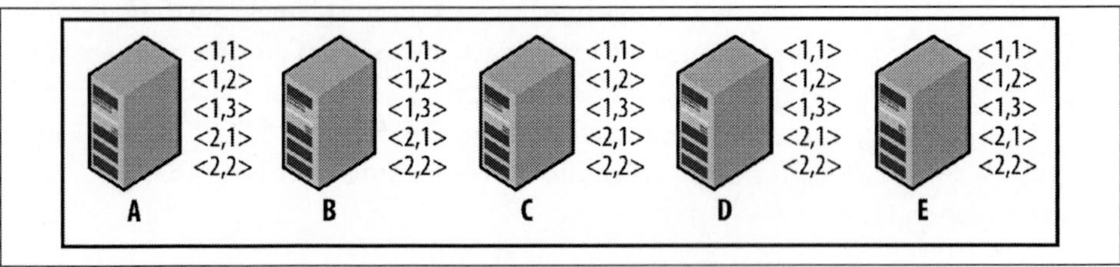


*Figure 10-3. An ensemble of five servers that has lost data*

This result is very bad. We have lost state, and the state of the replicas is no longer consistent with clients that observed ⟨1,4⟩, ⟨1,5⟩, and ⟨1,6⟩. To remedy this, ZooKeeper has a reconfigure operation. This means that administrators do not have to do the reconfiguration procedure by hand and risk corrupting the state. Even better, we don't even have to bring anything down.

Reconfiguration allows us to change not only the members of the ensemble, but also their network parameters. Because the configuration can change, ZooKeeper needs to move the reconfigurable parameters out of the static configuration file and into a configuration file that will be updated automatically. The `dynamicConfigFile` parameter links the two files together.

### Can I Use Dynamic Configuration?

This feature is currently only available in the trunk branch of the Apache repository. The target release for trunk is 3.5.0, although nothing really guarantees that the release number will be this one; it might change depending on how trunk evolves. The latest release at the time this book is being written is 3.4.5, and it does not include this feature.

Let's take an example configuration file that we have been using before dynamic configuration:

```
tickTime=2000
initLimit=10
syncLimit=5
dataDir=./data
dataLogDir=./txnlog
clientPort=2182
server.1=127.0.0.1:2222:2223
server.2=127.0.0.1:3333:3334
server.3=127.0.0.1:4444:4445
```

and change it to a configuration file that supports dynamic configuration:

```
tickTime=2000
initLimit=10
syncLimit=5
dataDir=./data
dataLogDir=./txnlog
dynamicConfigFile=./dyn.cfg
```

Notice that we have even removed the `clientPort` parameter from the configuration file. The *dyn.cfg* file is now going to be made up of just the server entries. We are adding a bit of information, though. Now each entry will have the form:

```
server.id=host:n:n[:role];[client_address:]client_port
```

Just as in the normal configuration file, the hostname and ports used for quorum and leader election messages are listed for each server. The `role` must be either `participant` or `observer`. If `role` is omitted, `participant` is the default. We also specify the `client_port` (the server port to which clients will connect) and optionally the address of a specific interface on that server. Because we removed `clientPort` from the static config file, we need to add it here.

So now our *dyn.cfg* file looks like this:

```
server.1=127.0.0.1:2222:2223:participant;2181
server.2=127.0.0.1:3333:3334:participant;2182
server.3=127.0.0.1:4444:4445:participant;2183
```

These files have to be created before we can use reconfiguration. Once they are in place, we can reconfigure an ensemble using the `reconfig` operation. This operation can operate incrementally or as a complete (bulk) update.

An incremental reconfig can take two lists: the list of servers to remove and the list of server entries to add. The list of servers to remove is simply a comma-separated list of server IDs. The list of server entries to add is a comma-separated list of server entries of the form found in the dynamic configuration file. For example:

```
reconfig -remove 2,3 -add \
  server.4=127.0.0.1:5555:5556:participant;2184,\
  server.5=127.0.0.1:6666:6667:participant;2185
```

This command removes servers 2 and 3 and adds servers 4 and 5. There are some conditions that must be satisfied in order for this operation to succeed. First, like with all other ZooKeeper operations, a quorum in the original configuration must be active. Second, a quorum in the new configuration must also be active.

**Reconfiguring from One to Many**

When we have a single ZooKeeper server, the server runs in standalone mode. This makes things a bit more complicated because a reconfiguration that adds servers not only changes the composition of quorums, but also switches the original server from standalone to quorum mode. At this time we have opted for not allowing reconfiguration for standalone deployments, so to use this feature you will need to start with a configuration in quorum mode.

ZooKeeper allows only one configuration change to happen at a time. Of course, the configuration operation happens very fast, and reconfiguration is infrequent enough that concurrent reconfigurations should not be a problem.

The *-file* parameter can also be used to do a bulk update using a new membership file. For example, *reconfig -file newconf* would produce the same result as the incremental operation if *newconf* contained:

```
server.1=127.0.0.1:2222:2223:participant;2181
server.4=127.0.0.1:5555:5556:participant;2184
server.5=127.0.0.1:6666:6667:participant;2185
```

The `-members` parameter followed by a list of server entries can be used instead of `-file` for a bulk update.

Finally, all the forms of `reconfig` can be made conditional. If the `-v` parameter is used, followed by the configuration version number, the `reconfig` will succeed only if the configuration is at the current version when it executes. You can get the version number of the current configuration by reading the `/zookeeper/config` znode or using the `config` command in *zkCli*.

**Manual Reconfiguration**

If you really want to do reconfiguration manually (perhaps you are using an older version of ZooKeeper), the easiest and safest way to do it is to make one change at a time and bring the ensemble all the way up (i.e., let a leader get established) and down between each change.

## Managing Client Connect Strings

We have been talking about ZooKeeper server configuration, but the clients have a bit of related configuration as well: the connect string. The client connect string is usually represented as a series of `host:port` pairs separated by a comma. The `host` can be specified either as an IP address or as a hostname. Using a hostname allows for a layer of indirection between the actual IP address of the server and the identifier used to access the server. It allows, for example, an administrator to replace a ZooKeeper server with a different one without changing the setup of the clients.

However, this flexibility is limited. The administrator can change the machines that make up the cluster, but not the machines being used by the clients. For example, in Figure 10-4, the ZooKeeper ensemble can be easily changed from a three-server ensemble to a five-server ensemble using reconfiguration, but the clients will still be using three servers, not all five.

There is another way to make ZooKeeper more elastic with respect to the number of servers, without changing client configuration. It is natural to think of a hostname resolving to a single IP address, but in reality a hostname can resolve to multiple addresses. If a hostname resolves to multiple IP addresses, the ZooKeeper client can connect to any of these addresses. In Figure 10-4, suppose the three individual IP addresses, `zk-a`, `zk-b`, and `zk-c`, resolved to 10.0.0.1, 10.0.0.2, and 10.0.0.3. Now suppose instead you use DNS to configure a single hostname, `zk`, to resolve to all three IP addresses. You can just change the number of addresses to five in DNS, and any client that subsequently starts will be able to connect to all five servers, as shown in Figure 10-5.

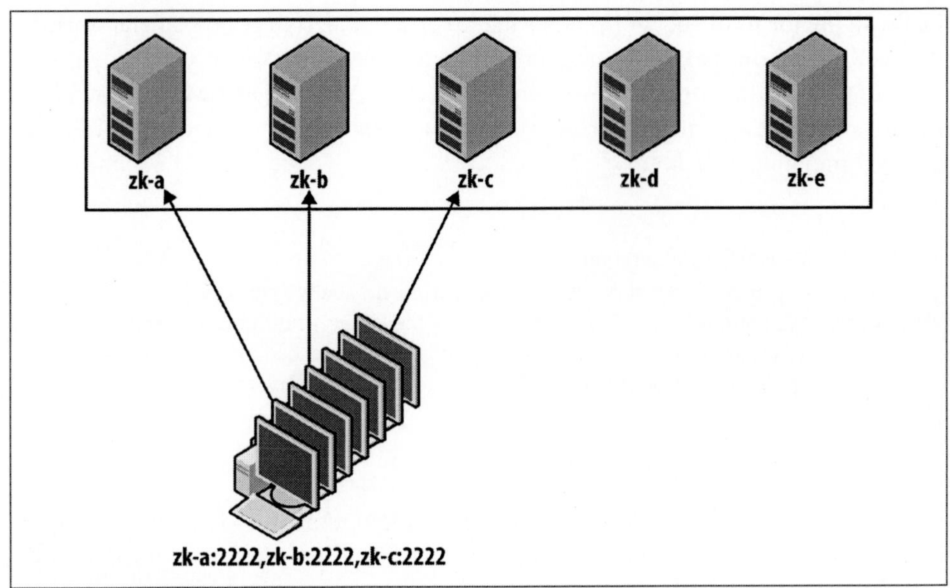


*Figure 10-4. Reconfiguring clients from using three servers to using five*

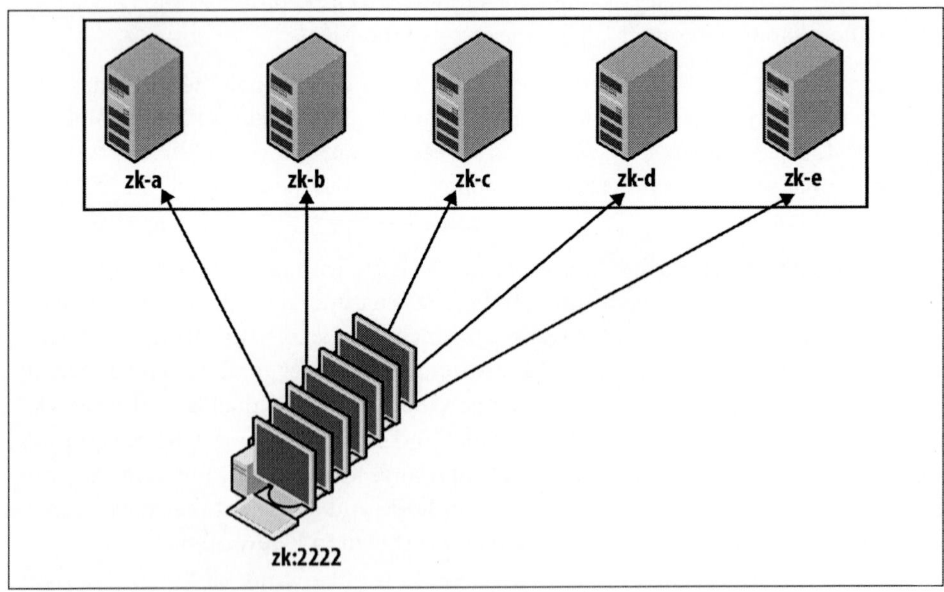


*Figure 10-5. Reconfiguring clients from using three servers to using five with DNS*

There are a couple of caveats to using a hostname that resolves to multiple addresses. First, all of the servers have to use the same client port. Second, hostname resolution currently happens only on handle creation, so clients that have already started do not recognize the new name resolution. It applies only to newly created ZooKeeper clients.

The client connect string can also include a path component. This path indicates the root to use when resolving pathnames. The behavior is similar to the `chroot` command in Unix, and you will often hear this feature referred to as "chroot" in the ZooKeeper community. For example, if a client specifies the connection string `zk:2222/app/superApp` when connecting and issues `getData("/a.dat", . . .)`, the client will receive the data from the znode at path `/app/superApp/a.dat`. (Note that there must be a znode at the specifed path. The connect string will not create one for you.)

The motivation for using a path component in a connect string is to allow a single ZooKeeper ensemble to host multiple applications without requiring them to append a prefix to all their paths. Each application can use ZooKeeper as if it were the only application using the ensemble, and administrators can carve up the namespace as they wish. Figure 10-6 shows examples of different connect strings that can be used to root client applications at different points in the tree.

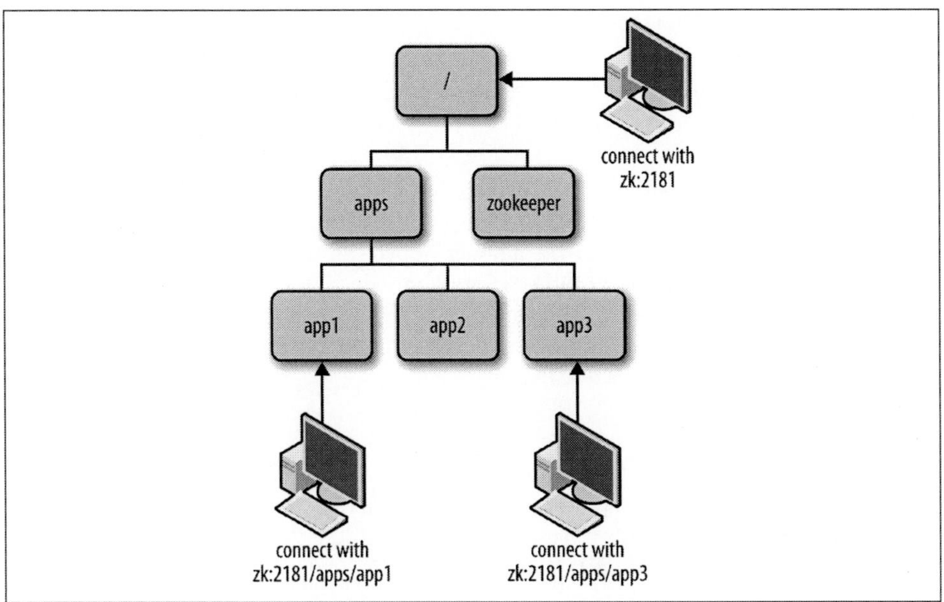


*Figure 10-6. Using the connect string to root ZooKeeper client handles*

**Overlapping Connection String**

When managing client connection strings, take care that a client connection string never includes hosts from two different ensembles. It's the quickest and easiest path to split brain.

# Quotas

Another configurable aspect of ZooKeeper is quotas. ZooKeeper has initial support for quotas on the number of znodes and the amount of data stored. It allows us to specify quotas based on a subtree and it will track the usage of the subtree. If a subtree exceeds its quota, a warning will be logged, but the operation will still be allowed to continue. At this point ZooKeeper detects when a quota is exceeded, but does not prevent processes from doing so.

Quota tracking is done in the special `/zookeeper` subtree. Applications should not store their own data in this subtree; instead, it is reserved for ZooKeeper's use. The `/zookeeper/quota` znode is an example of this use. To create a quota for the application `/application/superApp`, create the znode `/application/superApp` with two children: `zookeeper_limits` and `zookeeper_stats`.

The limit on the number of znodes is called the `count`, whereas the limit on the amount of data is called the `bytes`. The quotas for both `zookeeper_limits` and `zookeeper_stats` are specified as `count=`*`n`*`,bytes=`*`m`*, where *`n`* and *`m`* are integers. In the case of `zookeeper_limits`, *`n`* and *`m`* represent levels at which warnings will be triggered. (If one of them is `-1`, it will not act as a trigger.) In the case of `zookeeper_stats`, *`n`* and *`m`* represent the current number of znodes in the subtree and the current number of bytes in the data of the znodes of the subtree.

**Quota Tracking of Metadata**

The quota tracking for the number of bytes in the subtree does not include the metadata overhead for each znode. This metadata is on the order of 100 bytes, so if the amount of data in each znode is small, it is more useful to track the count of the number of znodes rather than the number of bytes of znode data.

Let's use *zkCli* to create `/application/superApp` and then set a quota:

```
[zk: localhost:2181(CONNECTED) 2] create /application ""
Created /application
[zk: localhost:2181(CONNECTED) 3] create /application/superApp super
Created /application/superApp
[zk: localhost:2181(CONNECTED) 4] setquota -b 10 /application/superApp
Comment: the parts are option -b val 10 path /application/superApp
[zk: localhost:2181(CONNECTED) 5] listquota /application/superApp
absolute path is /zookeeper/quota/application/superApp/zookeeper_limits
Output quota for /application/superApp count=-1,bytes=10
Output stat for /application/superApp count=1,bytes=5
```

We create `/application/superApp` with 5 bytes of data (the word "super"). We then set the quota on `/application/superApp` to be 10 bytes. When we list the quota on `/application/superApp` we see that we have 5 bytes left of our data quota and that we don't have a quota for the number of znodes under this subtree, because `count` is `-1` for the quota.

If we issue *get /zookeeper/quota/application/superApp/zookeeper_stats*, we can access this data directly without using the quota commands in *zkCli*. As a matter of fact, we can create and delete these files to create and delete quotas ourselves. If we run the following command:

```
create /application/superApp/lotsOfData ThisIsALotOfData
```

we should see the following entry in our log:

```
Quota exceeded: /application/superApp bytes=21 limit=10
```

# Multitenancy

Quotas, some of the throttling configuration options, and ACLs all make it worth considering the use ZooKeeper to host multiple tenants. There are some compelling reasons to do this:

- To provide reliable service, ZooKeeper servers should run on dedicated hardware. Sharing that hardware across multiple applications makes it easier to justify the capital investment.
- We have found that, for the most part, ZooKeeper traffic is extremely bursty: there are bursts of configuration or state changes that cause a lot of load followed by long periods of inactivity. If the activity bursts between applications are not correlated, making them share a server will better utilize hardware resources. Also remember to account for spikes generated when disconnect events happen. Some poorly written apps may generate more load than needed when processing a `Disconnected` event.
- By pooling hardware, we can achieve greater fault tolerance: if two applications that previously had their own ensembles of three servers are moved to a single cluster of five servers, they use fewer servers in total but can survive two server failures rather than just one.

When hosting multiple tenants, administrators will usually divide up the data tree into subtrees, each dedicated to a certain application. Developers can design their applications to take into account that their znodes need to have a prefix, but there is an easier way to isolate applications: using the path component in the connect string, described in "Managing Client Connect Strings" on page 197. Each application developer can write her application as if she has a dedicated ZooKeeper service. Then, if the administrator decides that the application should be deployed under `/application/newapp`, the application can use `host:port/application/newapp` rather than just `host:port`, and it will appear to the application that it is using a dedicated service. In the meantime, the administrator can set up the quota for `/application/newapp` to also track the space usage of the application.

# File System Layout and Formats

We have talked about snapshots and transaction logs and storage devices. This section describes how this is all laid out on the file system. A good understanding of the concepts discussed in "Local Storage" on page 170 will come in handy in this section; be prepared to refer back to it.

As we have already discussed, data is stored in two ways: transaction logs and snapshots. Both of these end up as normal files in the local file system. Transaction logs are written during the critical path of transaction processing, so we highly recommend storing them on a dedicated device. (We realize we have said this multiple times, but it really is important for good, consistent throughput and latency.) Not using a dedicated device for the transaction log does not cause any correctness issues, but it does affect performance. In a virtualized environment, for example, dedicated devices might not be available. Snapshots need not be stored on a dedicated device because they are taken lazily in a background thread.

Snapshots are written to the path specified in the `DataDir` parameter, and transaction logs are written to the path specified in the `DataLogDir` parameter. Let's take a look at the files in the transaction log directory first. If you list the contents of that directory, you will see a single subdirectory called *version-2*. We have had only one major change to the format of logs and snapshots, and when we made that change we realized that it would be useful to separate data by file versions to more easily handle data migration between versions.

## Transaction Logs

Let's look at a directory where we have been running some small tests, so it has only two transaction logs:

```
-rw-r--r--  1 breed 67108880 Jun  5 22:12 log.100000001
-rw-r--r--  1 breed 67108880 Jul 15 21:37 log.200000001
```

We can make a couple of observations about these files. First, they are quite large (over 6 MB each), considering the tests were small. Second, they have a large number as the filename suffix.

ZooKeeper preallocates files in rather large chunks to avoid the metadata management overhead of growing the file with each write. If you do a hex dump of one of these files, you will see that it is full of bytes of the null character (*\0*), except for a bit of binary data at the beginning. As the servers run longer, the null characters will be replaced with log data.

The log files contain transactions tagged with zxids, but to ease recovery and allow for quick lookup, each log file's suffix is the first zxid of the log file in hexadecimal. One nice thing about representing the zxid in hex is that you can easily distinguish the epoch part of the zxid from the counter. So, the first file in the preceding example is from epoch 1 and the second is from epoch 2.

Of course, it would be nice to be able to see what is in the files. This can be super useful for problem determination. There have been times when developers have sworn up and down that ZooKeeper is losing track of their znodes, only to find out by looking at the transaction logs that a client has actually deleted some.

We can look at the second log file with the following command:

```
java -cp $ZK_LIBS org.apache.zookeeper.server.LogFormatter version-2 /
log.200000001
```

This command outputs the following:

```
7/15/13... session 0x13...00 cxid 0x0 zxid 0x200000001 createSession 30000
7/15/13... session 0x13...00 cxid 0x2 zxid 0x200000002 create
'/test,#22746573746 ...
7/15/13... session 0x13...00 cxid 0x3 zxid 0x200000003 create
'/test/c1,#6368696c ...
7/15/13... session 0x13...00 cxid 0x4 zxid 0x200000004 create
'/test/c2,#6368696c ...
7/15/13... session 0x13...00 cxid 0x5 zxid 0x200000005 create
'/test/c3,#6368696c ...
7/15/13... session 0x13...00 cxid 0x0 zxid 0x200000006 closeSession null
```

Each transaction in the log file is output as its own line in human-readable form. Because only change operations result in a transaction, you will not see read transactions in the transaction log.

## Snapshots

The naming scheme for snapshot files is similar to the transaction log scheme. Here is the list of snapshots on the server used earlier:

```
-rw-r--r--  1 br33d  296 Jun  5 07:49 snapshot.0
-rw-r--r--  1 br33d  415 Jul 15 21:33 snapshot.100000009
```

Snapshot files are not preallocated, so the size more accurately reflects the amount of data they contain. The suffix used reflects the current zxid when the snapshot started. As we discussed earlier, the snapshot file is actually a fuzzy snapshot; it is not a valid snapshot on its own until the transaction log is replayed over it. Specifically, to restore a system, you must start replaying the transaction log starting at the zxid of the snapshot suffix or earlier.

The files themselves also store the fuzzy snapshot data in binary form. Consequently, there is another tool to examine the snapshot files:

```
java -cp ZK_LIBS org.apache.zookeeper.server.SnapshotFormatter version-2 /
snapshot.100000009
```

This command outputs the following:

```
 ----
/
  cZxid = 0x00000000000000
  ctime = Wed Dec 31 16:00:00 PST 1969
  mZxid = 0x00000000000000
  mtime = Wed Dec 31 16:00:00 PST 1969
  pZxid = 0x00000100000002
  cversion = 1
  dataVersion = 0
  aclVersion = 0
  ephemeralOwner = 0x00000000000000
  dataLength = 0
 ----
/sasd
  cZxid = 0x00000100000002
  ctime = Wed Jun 05 07:50:56 PDT 2013
  mZxid = 0x00000100000002
  mtime = Wed Jun 05 07:50:56 PDT 2013
  pZxid = 0x00000100000002
  cversion = 0
  dataVersion = 0
  aclVersion = 0
  ephemeralOwner = 0x00000000000000
  dataLength = 3
 ----
....
```

Only the metadata for each znode is dumped. This allows an administrator to figure out things like when a znode was changed and which znodes are taking up a lot of memory. Unfortunately, the data and ACLs don't appear in this output. Also, remember when doing problem determination that the information from the snapshot must be merged with the information from the log to figure out what was going on.

## Epoch Files

Two more small files make up the persistent state of ZooKeeper. They are the two epoch files, named *acceptedEpoch* and *currentEpoch*. We have talked about the notion of an epoch before, and these two files reflect the epoch numbers that the given server process has seen and participated in. Although the files don't contain any application data, they are important for data consistency, so if you are doing a backup of the raw data files of a ZooKeeper server, don't forget to include these two files.

## Using Stored ZooKeeper Data

One of the nice things about ZooKeeper data is that the standalone servers and the ensemble of servers store data in the same way. We've just mentioned that to get an accurate view of the data you need to merge the logs and the snapshot. You can do this by copying log files and snapshot files to another machine (like your laptop, for example), putting them in the empty *data* directory of a standalone server, and starting the server. The server will now reflect the state of the server that the files were copied from. This technique allows you to capture the state of a server in production for later review.

This also means that you can easily back up a ZooKeeper server by simply copying its data files. There are a couple of things to keep in mind if you choose to do this. First, ZooKeeper is a replicated service, so there is redundancy built into the system. If you do take a backup, you need to back up only the data of one of the servers.

It is important to keep in mind that when a ZooKeeper server acknowledges a transaction, it promises to remember the state from that time forward. So if you restore a server's state using an older backup, you have caused the server to violate its promise. This might not be a big deal if you have just suffered data loss on all servers, but if you have a working ensemble and you move a server to an older state, you risk causing other servers to also start to forget things.

If you are recovering from data loss on all or a majority of your servers, the best thing to do is to grab your latest captured state (from a backup from the most up-to-date surviving server) and copy that state to all the other servers before starting any of them.

# Four-Letter Words

Now that we have our server configured and up and running, we need to monitor it. That is where four-letter words come in. We have already seen some examples of four-letter words when we used *telnet* to see the system status in "Running the Watcher Example" on page 49. Four-letter words provide a simple way to do various checks on the system. The main goal with four-letter words is to provide a very simple protocol that can be used with simple tools, such as *telnet* and *nc*, to check system health and diagnose problems. To keep things simple, the output of a four-letter word will be human readable. This makes the words easy to experiment with and use.

It is also easy to add new words to the server, so the list has been growing. In this section we will point out some of the commonly used words. Consult the ZooKeeper documentation for the most recent and complete list of words:

`ruok`
: Provides (limited) information about the status of the server. If the server is running, it will respond with `imok`. It turns out that "OK" is a relative concept, though. For example, the server might be running but unable to communicate with the other servers in the ensemble, yet still report that it is "OK." For a more detailed and reliable health check, use the `stat` word.

`stat`
: Provides information about the status of the server and the connections that are currently active. The status includes some basic statistics and whether the server is currently active, if it is a leader or follower, and the last zxid the server has seen. Some of the statistics are cumulative and can be reset using the `srst` word.

`srvr`
: Provides the same information as `stat`, except the connection information, which it omits.

`dump`
: Provides session information, listing the currently active sessions and when they will expire. This word can be used only on a server that is acting as the leader.

`conf`
: Lists the basic server configuration parameters that the server was started with.

`envi`
: Lists various Java environment parameters.

`mntr`
: Offers more detailed statistics than `stat` about the server. Each line of output has the format `key<tab>value`. (The leader will list some additional parameters that apply only to leaders.)

`wchs`
: Lists a brief summary of the watches tracked by the server.

`wchc`
: Lists detailed information on the watches tracked by the server, grouped by session.

`wchp`
: Lists detailed information on the watches tracked by the server, grouped by the znode path being watched.

`cons`, `crst`
: `cons` lists detailed statistics for each connection on a server and `crst` resets all the connection counters to zero.((("cons"))

## Monitoring with JMX

Four-letter words are great for monitoring, but they do not provide a way to control or make changes to the system. ZooKeeper also uses a standard Java management protocol called JMX (Java Management Extensions) to provide more powerful monitoring and management capabilities. There are many books about how to set up and use JMX and many tools for managing servers with JMX; in this section we will use a simple management console called *jconsole* to explore the ZooKeeper management functionality that is available via JMX.

The *jconsole* utility is distributed with Java. In practice, a JMX tool such as *jconsole* is used to monitor remote ZooKeeper servers, but for this part of the exercise we will run it on the same machine as our ZooKeeper servers.

First, let's start up the second ZooKeeper server (the one with the ID of 2). Then we'll start *jconsole* by simply running the *jconsole* command from the command line. You should see a window similar to Figure 10-7 when *jconsole* starts.

Notice the process that has "zookeeper" in the name. This is the local process that *jconsole* has discovered that it can connect to.

Now let's connect to the ZooKeeper process by double-clicking that process in the list. We will be asked about connecting insecurely because we do not have SSL set up. Clicking the insecure connection button should bring up the screen shown in Figure 10-8.

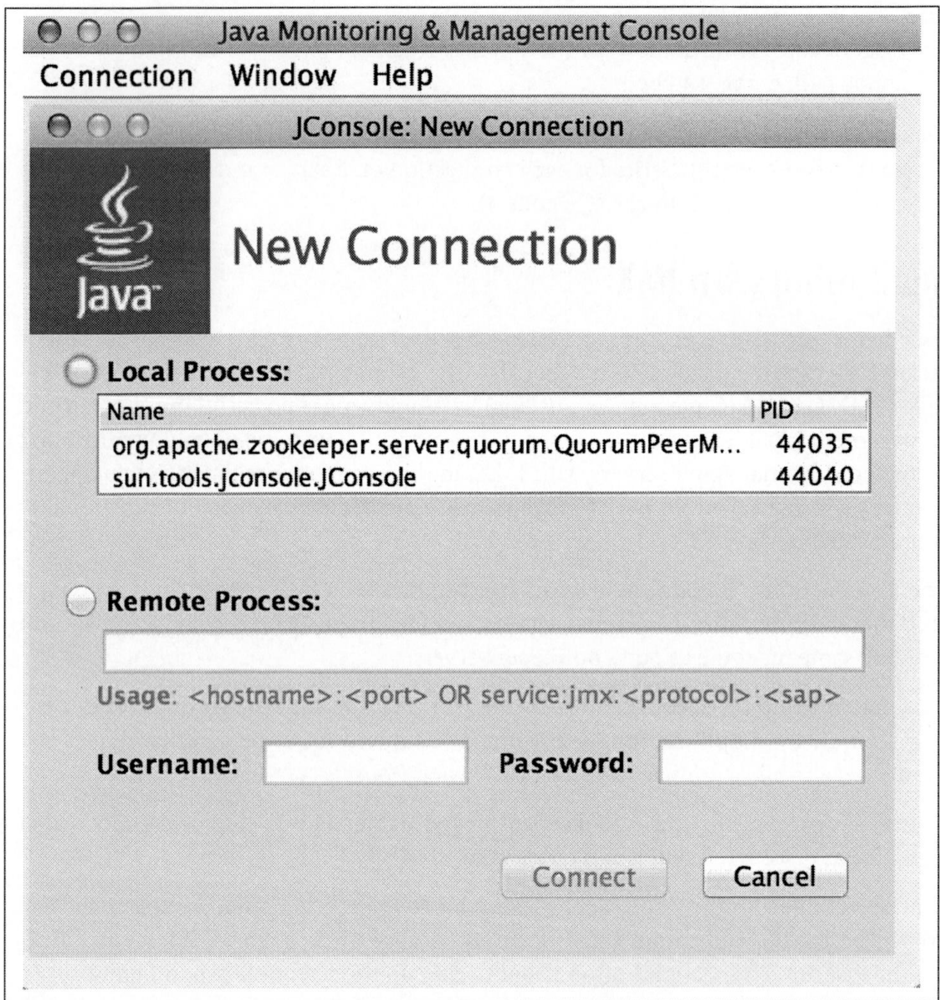


*Figure 10-7. jconsole startup screen*

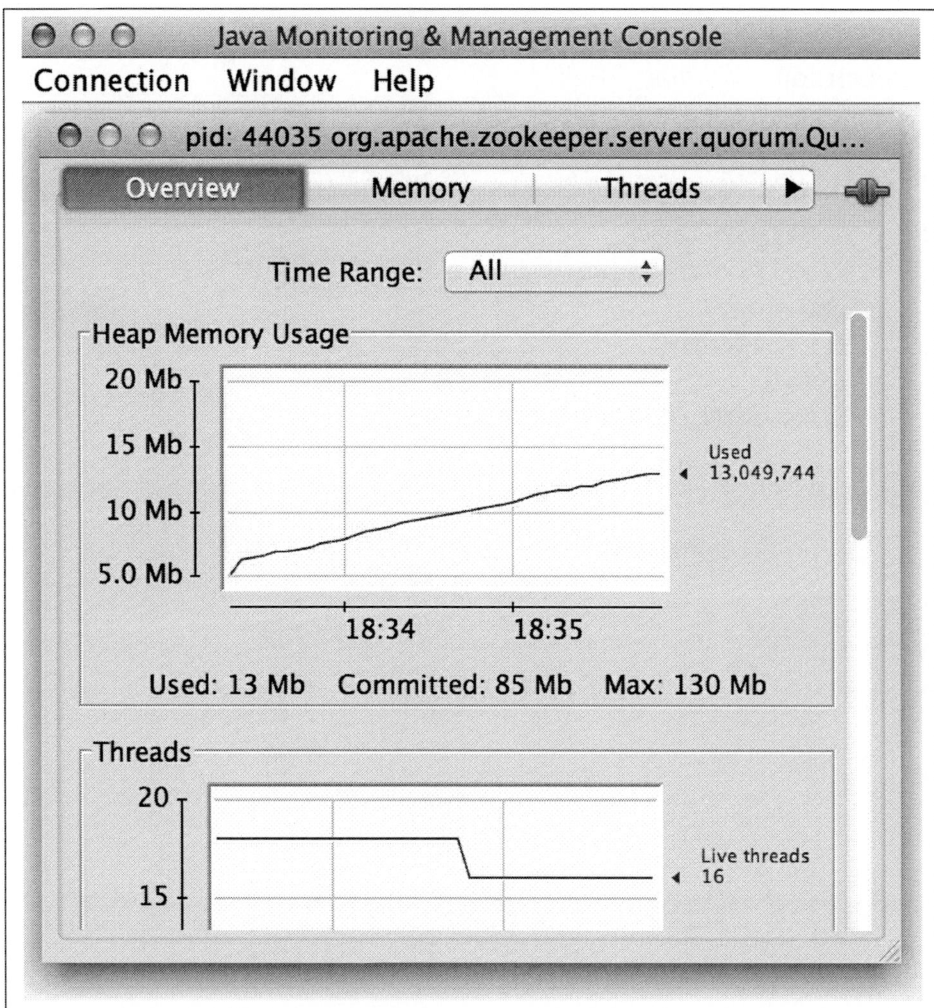


*Figure 10-8. The first management window for a process*

As we can see from this screen, we can get various interesting statistics about the ZooKeeper server process with this tool. JMX allows customized information to be exposed to remote managers through the use of MBeans (Managed Beans). Although the name sounds goofy, it is a very flexible way to expose information and operations. *jconsole* lists all the MBeans exposed by the process in the rightmost information tab, as shown in Figure 10-9.

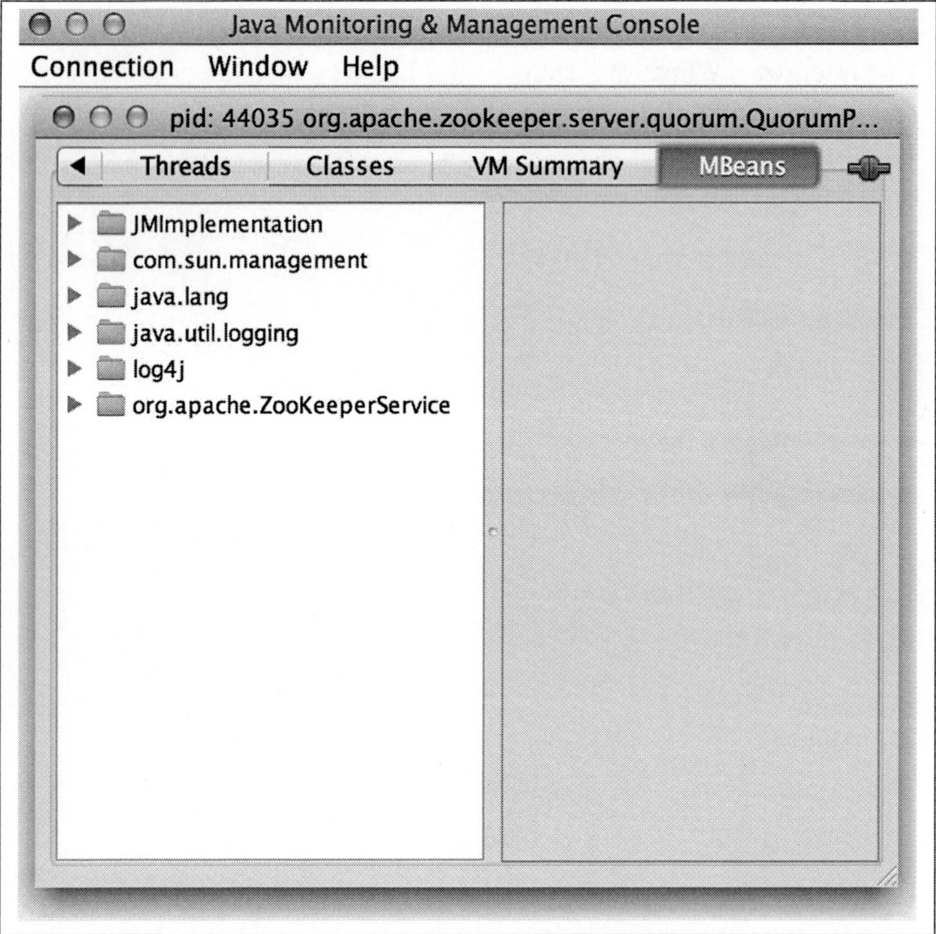


*Figure 10-9. jconsole MBeans*

As we can see from the list of MBeans, some of the components used by ZooKeeper are also exposed via MBeans. We are interested in the `ZooKeeperService`, so we will double-click on that list item. We will see a hierarchal list of replicas and information about those replicas. If we open some of the subentries in the list, we will see something like Figure 10-10.

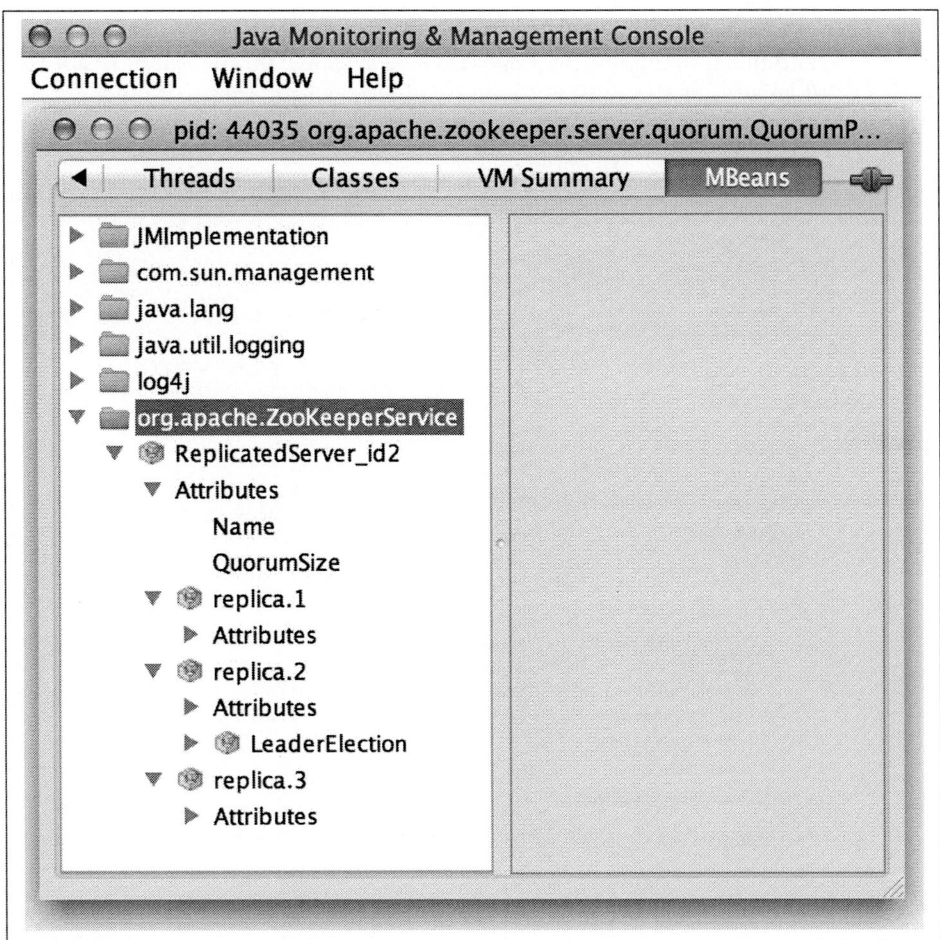


*Figure 10-10. jconsole information for server 2*

As we explore the information for *replica.2* we will notice that it also includes some information about the other replicas, but it's really just the contact information. Because server 2 doesn't know much about the other replicas, there is not much more it can reveal about them. Server 2 does know a lot about itself, though, so it seems like there should be more information that it can expose.

If we start up server 1 so that server 2 can form a quorum with server 1, we will see that we get more information about server 2. Start up server 1 and then check server 2 in *jconsole* again. Figure 10-11 shows some of the additional information that is exposed by JMX. We can now see that server 2 is acting as a follower. We can also see information about the data tree.

Figure 10-11 shows the JMX information for server 1. As we see, server 1 is acting as a leader. One additional operation, `FollowerInfo`, is available on the leader to list the followers. When we click this button, we see a rather raw list of information about the other ZooKeeper servers connected to server 1.

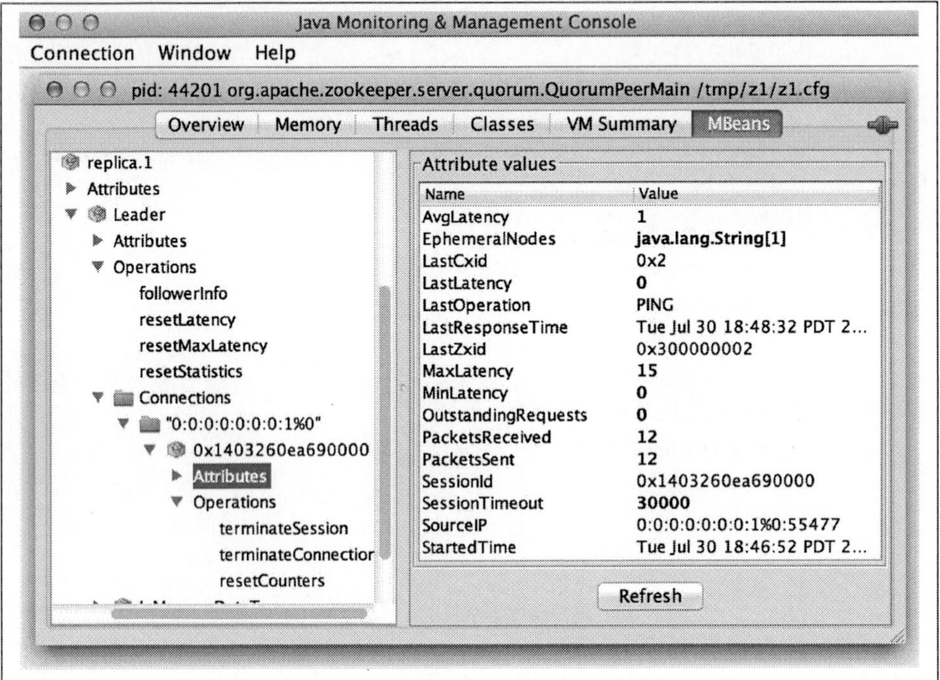


*Figure 10-11. jconsole information for server 1*

Up to now, the information we've seen from JMX looks prettier than the information we get from four-letter words, but we really haven't seen any new functionality. Let's look at something we can do with JMX that we cannot do with four-letter words. Start a *zkCli* shell. Connect to server 1, then run the following command:

```
create -e /me "foo"
```

This will create an ephemeral znode on the server. Figure 10-11 shows that a new informational entry for Connections has appeared in the JMX information for server 1. The attributes of the connection list various pieces of information that are useful for debugging operational issues. This view also exposes two interesting operations: `terminateSession` and `terminateConnection`.

The `terminateConnection` operation will close the ZooKeeper client's connection to the server. The session will still be active, so the client will be able to reconnect to another server; the client will see a disconnection event but should be able to easily recover from it.

In contrast, the `terminateConnection` operation declares the session dead. The client's connection with the server will close and the session will be terminated as if it has expired. The client will not be able to connect to another server using the session. Care should be taken when using `terminateConnection` because that operation can cause the session to expire long before the session timeout, so other processes may find out that the session is dead before the process that owns that session finds out.

## Connecting Remotely

The JMX agent that runs inside of the JVM of a ZooKeeper server must be configured properly to support remote connections. There are a variety of options to configure remote connections for JMX. In this section we show one way of getting JMX set up to see what kind of functionality it provides. If you want to use JMX in production, you will probably want to use another JMX-specific reference to get some of the more advanced security features set up properly.

All of the JMX configuration is done using system properties. The *zkServer.sh* script that we use to start a ZooKeeper server has support for setting these properties using the `SERVER_JVMFLAGS` environment variable.

For example, we can access server 3 remotely using port 55555 if we start the server as follows:

```
SERVER_JVMFLAGS="-Dcom.sun.management.jmxremote.password.file=passwd \
  -Dcom.sun.management.jmxremote.port=55555 \
  -Dcom.sun.management.jmxremote.ssl=false \
  -Dcom.sun.management.jmxremote.access.file=access"
_path_to_zookeeper_/bin/zkServer.sh start _path_to_server3.cfg_
```

The properties refer to a password and access file. These have a very simple format. Create the *passwd* file with:

```
# user password
admin <password>
```

Note that the password is stored in clear text. For this reason, the password file must be readable and writable only by the owner of the file; if it is not, Java will not start up. Also, we have turned off SSL. That means the password will go over the network in clear text. If you need stronger security, there are much stronger options available to JMX, but they are outside the scope of this book.

For the access file, we are going to give `readwrite` privileges to `admin` by creating the file with:

```
admin readwrite
```

Now, if we start *jconsole* on another computer, we can use `host:5555` for the remote process location (where `host` is the hostname or address of the machine running ZooKeeper), and the user `admin` with the password `<password>` to connect. If you happen to misconfigure something, *jconsole* will fail with messages that give little clue about what is going on. Starting *jconsole* with the `-debug` option will provide more information about failures.

# Tools

Many tools and utilities come with ZooKeeper or are distributed separately. We have already mentioned the log formatting utilities and the JMX tool that comes with Java. In the *contrib* directory of the ZooKeeper distribution, you can find utilities to help integrate ZooKeeper into other monitoring systems. A few of the most popular offerings there are:

- Bindings for Perl and Python, implemented using the C binding.
- Utilities for visualizing the ZooKeeper logs.
- A web UI for browsing the cluster nodes and modifying ZooKeeper data.
- *zktreeutil*, which comes with ZooKeeper, and *guano*, which is available on GitHub. These utilities conveniently import and export data to and from ZooKeeper.
- `zktop`, also available on GitHub, which monitors ZooKeeper load and presents it in a Unix `top`-like interface.
- ZooKeeper Smoketest, available on GitHub. This is a simple smoketest client for a ZooKeeper ensemble; it's a great tool for developers getting familiar with ZooKeeper.

Of course, this isn't an exhaustive list, and many of the really great tools for running ZooKeeper are developed and distributed outside of the ZooKeeper distribution. If you are a ZooKeeper administrator, it would be worth your while to try out some of these tools in your environment.

# Takeaway Messages

Although ZooKeeper is simple to get going, there are many ways to tweak the service for your environment. ZooKeeper's reliability and performance also depend on correct configuration, so it is important to understand how ZooKeeper works and what the different parameters do. ZooKeeper can adjust to various network topologies if the timing and quorum configurations are set properly. Although changing the members of a ZooKeeper ensemble by hand is risky, it is a snap with the ZooKeeper `reconfig` operation. There are many tools available to make your job easier, so take a bit of time to explore what is out there.

# Index

## A



## B



## C



*We'd like to hear your suggestions for improving our indexes. Send email to index@oreilly.com.*

## D



## E

## F



## G



## H



## I



## J



## K



## L

## S

## About the Authors

**Flavio Junqueira** is a member of the research staff of Microsoft Research in Cambridge, UK. He holds a PhD degree in computer science from the University of California, San Diego. He is interested in various aspects of distributed systems, including distributed algorithms, concurrency, and scalability. He is an active contributor of Apache projects, such as Apache ZooKeeper (PMC chair and committer) and Apache BookKeeper (committer). When he is idle, he sleeps.

**Benjamin Reed** is a Software Engineer at Facebook working on all things small. His previous positions include Principal Research Scientist at Yahoo! Research (working on all things big) and Research Staff Member (working on the big and the small) at IBM Almaden Research. The University of California, Santa Cruz granted him a PhD in computer science. He has worked in the areas of distributed computing, big data processing, distributed storage, systems management, and embedded frameworks. He participated in various open source projects such as Hadoop and Linux. He helped start the Pig, ZooKeeper, and BookKeeper projects hosted by the Apache Software Foundation.

## Colophon

The animal on the cover of *ZooKeeper* is a European wildcat (*Felis silvestris silvestris*), a subspecies of the wildcat that inhabits the forests and grasslands of Europe, as well as Turkey and the Caucasus Mountains.

Similar in size to a large domestic cat, the European wildcat has a broader head, longish fur, and a shorter, blunted tail—white patches are often found on the throat, chest, and abdomen. The staple diet for the majority of European wildcats is made up of small rodents such as wood mice, pine voles, water voles, and shrews. Interestingly, at odds with domesticated cats' love of fish, wildcats rarely prey on fish in the wild.

The European wildcat was once found throughout Europe and is considered by some to be the oldest form of the species—limited fossil records indicate an ancestral link to wildcats dating back to the Early Pleistocene period. During the past 300 years, the range of the European wildcat, through pressures brought about by hunting and the spread of human population, has been significantly reduced.

Hybridization is also a major issue. Although many of the wildcat subspecies live in remote regions, others are in relative close proximity to human habitation and therefore near domestic and feral cat populations, within which they often mate. Over an extended period of time, it is possible that certain subspecies will simply "breed" themselves out of existence.

The cover image is from Meyers Kleines Lexicon. The cover fonts are URW Typewriter and Guardian Sans. The text font is Adobe Minion Pro; the heading font is Adobe Myriad Condensed; and the code font is Dalton Maag's Ubuntu Mono.